NORFOLK ORIGINS

CHANGING AGRICULTURE

Other titles in the Norfolk Origins series:

1: Hunters to First Farmers (published 1981)

2: Roads and Tracks (published 1983)

3: Celtic Fire and Roman Rule (1987; 2nd edition 2002)

4: The North Folk: Angles, Saxons & Danes (published 1990)

5: Deserted Villages in Norfolk (published 1996)

NORFOLK ORIGINS 6

Changing Agriculture
in Georgian and Victorian Norfolk

Susanna Wade Martins

POPPYLAND
PUBLISHING

Copyright © 2002 Susanna Wade Martins

ISBN 0 946148 58 9

Published by Poppyland Publishing, 4 Alfred Road, Cromer NR27 9AN

All rights reserved. No part of this publication may be reproduced, stored in a retrieval system or transmitted in any form by any means electronic, mechanical, photocopying, recording or otherwise, without the prior permission of the publishers.

Picture credits
Bodleian Library, University of Oxford (shelfmark 17363 b. 28): p. 102
Lady Hare: p. 78
Philip Judge: pp. 12, 34, 35, 41, 47 (line), 50
Benjamin Bulwer Long: pp. 17–19
Nicholas Meade: p. 15
Reproduced by permission of His Grace the Duke of Norfolk: p. 16
Norfolk Rural Life Museum, Norfolk Museums & Archaeology Service: pp. 87, 100, 101
Poppyland collection: p.88
Poppyland Photos: pp. 37, 38, 54, 86, 90
University of East Anglia, Centre of East Anglian Studies: pp. 20, 76, 79
Wolterton Hall archive: Cover Wolterton map extract

Designed and typeset in 9½ on 12 pt Arial by Watermark, Cromer NR27 9HL

Printed by Printing Services (Norwich) Ltd

Contents

Introduction	7
The origins of 'Norfolk agriculture' 1670–1720	10
The forces of change 1720–1840	29
Landlords and 'improvement'	29
The farmers	45
The labourers	62
High farming c.1840–1875	74
Norfolk farming in a changing world 1875–1914	89
Sources	105
Index	108

*The emblem of the
Royal Agricultural Society of England
(see page 9)*

INTRODUCTION

In the period covered by this book (1670–1914), Britain was transformed from an agricultural economy to an urban nation with most of its wealth made in factories. People were moving from the countryside to the towns, and it was also a period of massive population growth, so the rural population had to produce food for a large new industrial workforce. Up until the 1870s, nearly all this food had to be found from within the British Isles. This could not have been achieved without fundamental changes in production techniques. The agricultural developments of this period must be seen as the most important condition for the creation of the industrial society of the nineteenth century; but this was a two-way process. Industrial products and ideas as well as finance was making possible the modernisation of agriculture.

But it is easy to exaggerate the speed and importance of these changes. Before regular censuses began in 1801, there are no figures for the gradual shift of population, but it had begun well before then and was not complete until after 1914. As late as 1851 half the population still lived in the countryside and it was not until after that date that Norfolk villages ceased to grow. As late as 1901, agriculture was still the largest single employer in England and Wales. Just over one in ten (44,380 out of a total population of 476,553) were employed in agriculture in Norfolk.

It is also important to bear in mind that the contrasts between town and country, agricultural labourer and industrial worker, were not nearly so clear-cut before 1800 as they were to become later. As late as 1722, Daniel

CHANGING AGRICULTURE

Defoe, the author of *Robinson Crusoe,* saw Norfolk as an industrial county. He wrote in his diary of a tour through the eastern counties that spinning and weaving provided employment for women and children and, particularly in the north-east of the county, the wealth of villages such as Worstead was based on the textile industry rather than agriculture. Only gradually did industrial employment for the rural labourer decline. However, by 1834 a Royal Commission into the operation of the Poor Laws found little evidence for employment opportunities outside agriculture. There are only a few references to spinning and weaving. In Costessey, near Norwich, a witness reported that 'formerly the women had spinning to do, and brought in as much as the men. This I remember when I started farming.' In Scole 'spinning and weaving have been superseded by machinery and women and children have little to do except at harvest time'. It was in fact Norfolk's position as a de-industrialising society that meant a large cheap labour force was available to implement some of the agricultural changes which made Norfolk famous as the pioneer of agricultural change during this period.

It is clear that a massive shift in the economic base of Great Britain took place in the 200 years after 1670, and that it was a change in agricultural organisation and methods which allowed the country to become more industrial and urban. It has long been held that it was in East Anglia, and particularly in Norfolk, that these developments first took place. It was certainly here that they received most publicity: self-promoters such as Thomas William Coke at Holkham were followed by the first generation of agricultural historians, who tended to see all 'improvement' as the work of 'great men' (Prothero, *English Farming, Past and Present* (1912)). The first historian to write specifically about Norfolk was Naomi Riches. Her book *The Agricultural Revolution in Norfolk* was published in 1937 and her studies tended to follow in the footsteps of Prothero, mainly because of the sort of material that was available for her to study. Her researches were mainly confined to Holkham, using material in the University of Chicago library as well as many eighteenth and nineteenth century books and pamphlets on agriculture. Here she found plenty of evidence of 'improvement'.

In recent years, more evidence has come to light from other sources. Not only are the papers of smaller and less famous estates now finding their way into record offices, but other sources are being used. Inventories of the possessions and crops of farmers have been studied for what they can tell us about the practices of a whole range of working farmers. Tithe accounts list the crops grown in a parish on which tithes were payable to the incumbent. Map evidence may well indicate crops grown in fields and cropping

INTRODUCTION

patterns as well as the size of farms and the types of building required to work the farm, whilst surviving buildings themselves indicate the status of farmers and their farming methods. The development of the subject and techniques of landscape history have also allowed the farms, fields and hedgerows to be studied for evidence of previous farming systems. Some farmers (although admittedly probably not typical ones) kept diaries which can also be used. Parliamentary papers contain reports on agriculture and employment, and although the evidence contained in them is undoubtedly slanted to prove a point, they also contain valuable information. It is now possible, therefore, to revisit the subject of agricultural history, drawing on a much wider variety of sources than was possible in Riches' day, and consider farming development from evidence not associated with the great estates. Modern historians tend to lay more emphasis on the farmers themselves as innovators, and the varying roles of landlords, farmers and labourers in innovation and investment in agriculture will be one of the main themes of this book.

Another difference between historians writing now and those of an earlier generation is that we are no longer as confident about the inevitable march of progress as we once were. We do not any longer see history as a gradual unrolling of a carpet towards a better world. We are diffident about the use of the word 'improvement'. In this we are entirely different from the agriculturalists of the eighteenth and nineteenth centuries who wrote confident articles and books and founded the Royal Agricultural Society in 1839 with the motto 'Practice with Science'. This culture of improvement will be put into context as we consider whether all 'improvements' of the time were sustainable in the long run and whether some of the pillars of the 'Norfolk system' were indeed suited to conditions in other parts of the United Kingdom.

This book draws heavily on research carried out by the writer with Dr Tom Williamson at the Centre of East Anglian Studies, University of East Anglia. It will explore the origins of 'Norfolk agriculture' as understood by the improvers of the time and will emphasise the role of the farmer as well as the landlord. It will also consider the sustainability of such systems in the period of depressed farm prices after 1870, particularly those developed during the periods of optimism and high grain prices of the Napoleonic wars, and later those relying on expensive inputs during the years of intensive farming of the mid-nineteenth century known as 'high farming'.

THE ORIGINS OF 'NORFOLK AGRICULTURE' 1670–1720

What is 'Norfolk agriculture'?

There is no denying that the period covered by this book was one of enormous change. Agriculture moved from, in some areas, an almost medieval open-field system where two or more crops of grain were followed by a period of fallow and where areas of commons and heaths supported livestock for much of the year. This was replaced by enclosed fields where the soil was carefully manured, using marl, lime and animal manures, and draining by laying furze and hedge clippings (known as 'bush') in open drains which were then covered over, or later by using piped systems. The land could then be cultivated to produce higher yields from crops grown in rotations that eliminated the need for the fallow year.

The new crops which enabled these rotations to be implemented were 'artificial' grasses (imported varieties such as clover and sainfoin), and root crops such as turnips. Both of these provided fodder which allowed more animals to be kept and so more manure to be produced. Clover was valuable as a plant which restored nitrogen to the soil. Turnips had the advantage that, if they were planted in rows, or drilled in, then the ground could be weeded between the rows. The turnips could therefore be regarded as a cleaning plant, so reducing the need for a fallow year during which several ploughings were necessary to kill the weeds.

These developments were by the nineteenth century regarded as the main ingredients of 'Norfolk agriculture', which was seen as the epitome of improved husbandry and was transplanted wholesale around Britain. The standard of farming in any one region was judged by the degree to which this rigid formula was adhered to. The pace of its implementation has been the subject of much discussion over the years and has been the basis of debates about the timing of the 'agricultural revolution'. As we shall see, the pace and timing of change varied from region to region and it may be that developments that took over two generations to implement are too slow to be regarded as 'revolutionary' anyway.

Whilst most of these indicators of improved agriculture were evident in Norfolk agriculture by 1840, there were many other changes still to come. Some historians see the years 1840–1875 as demonstrating far more dramatic changes than those before. In 1968, the agricultural historian, Professor Thompson went so far as to describe the period as the 'second agricultural revolution'. Unlike the earlier developments, they are not particularly associated with Norfolk, nor can regional variations be identified within the county, but they were as significant for Norfolk agriculture as elsewhere and will be considered in a later chapter.

The agricultural regions of seventeenth-century Norfolk

The Norfolk of 1670 would have looked very different from that we know today, and, the regional differences across the county would have been particularly striking. The gradual breaking down of these regional characteristics is one of the themes of this book. Much of this variety was based on the soil regions of the county, which were mapped for the agricultural journalist Arthur Young's *General View of the Agriculture of the County of Norfolk* as early as 1804. Much of his classification is still accepted today. The soils were deposited by the retreating glaciers at the end of the last ice age, and so there can be a bewildering variety of soils even within one field. This means that although generalisations about soil regions are helpful, we need to be aware that there can be great variations, even in one supposedly homogenous area.

The basic geology of the county is influenced by a deposit of stratified sedimentary rocks, the most important of which is chalk, which was laid down and then tilted so that the chalk reaches the surface in the north-west of the county and is exposed as cliffs at Hunstanton. It then dips towards the south east, so that while the chalk is not far from the surface in the north-

The soils of Norfolk

Peat and silt
Light loams
Medium clays
Heavy clays
Acid sands and gravels

0　5　10　15 km

west, it is buried ever deeper under glacial deposits southwards and eastwards. The nearer the chalk is to the surface, the better drained will be the soils. Immediately behind the chalk escarpment are the 'good sands region' first identified by Arthur Young. They consist of a dark loamy topsoil over a strongly weathered clayey subsoil. This is a versatile land which today is suited to both arable and grassland.

Across the centre of the county stretches the great arc of a boulder clay plateau, whilst in the north-east are light and fertile loams. Much of the boulder clay is heavy and difficult to work, although there are pockets of lighter soils within it. These soils for the most part need draining before they can be used for extensive arable farming.

The lighter eastern soils are richest to the south where they border onto the flooded estuarine peats of the Norfolk Broads. They form some of the most fertile farmland in the country and have supported a large and wealthy agricultural population, as the density of market towns and villages, with fine medieval churches and Tudor manor houses, testifies.

To the north, the sands and gravels of the glacial moraine which forms the Cromer Ridge separate this area from the boulder clay and chalk to the north and form areas of infertile heath penetrating deep into this otherwise productive area. To the west, between the chalk scarp and the fens is an area of mixed, mostly sandy soils in which the only good building stone in the county, known as carstone, is found.

In great contrast to the heavy claylands of much of Norfolk are the light sands of Breckland. These sandy soils laid on chalk are extremely dry and difficult to cultivate, with the only good soil along the river valleys, where the villages are to be found. Here, over-grazing in the middle ages left land bare and most suited to rabbit warrens. It was not until the late eighteenth century that efforts were made to bring some of the drier parts into cultivation.

Finally, there are the areas of marsh in the broads and the fens. Drainage here began in the seventeenth century, but it was not until the nineteenth that they were fully reclaimed. The fens were mainly used for arable, while the broads remained an important grazing area.

◄ *The soils of Norfolk vary from the heaviest of clays in the south to the lightest of sands in the west. The best are the light loams of the east and the chalks of the north-west, but everywhere local glacial deposits of clays, sands and gravel mean that there can be great variations in soil quality within a single farm.*

CHANGING AGRICULTURE

It is against this varied geological and soil background that the farming regions of the seventeenth century emerged.

The claylands

This area represents the largest of the agricultural regions in Norfolk and typically consisted by the seventeenth century of dispersed settlement connected by a maze of small lanes, with parish churches either on the periphery of settlements or left isolated in the middle of fields as their medieval populations moved on, often to the edges of greens. Although the area would no doubt have been mostly open fields in the middle ages, many of these had been enclosed by the end of the seventeenth century into small, irregular closes surrounded by thick hedges, rich in standard and

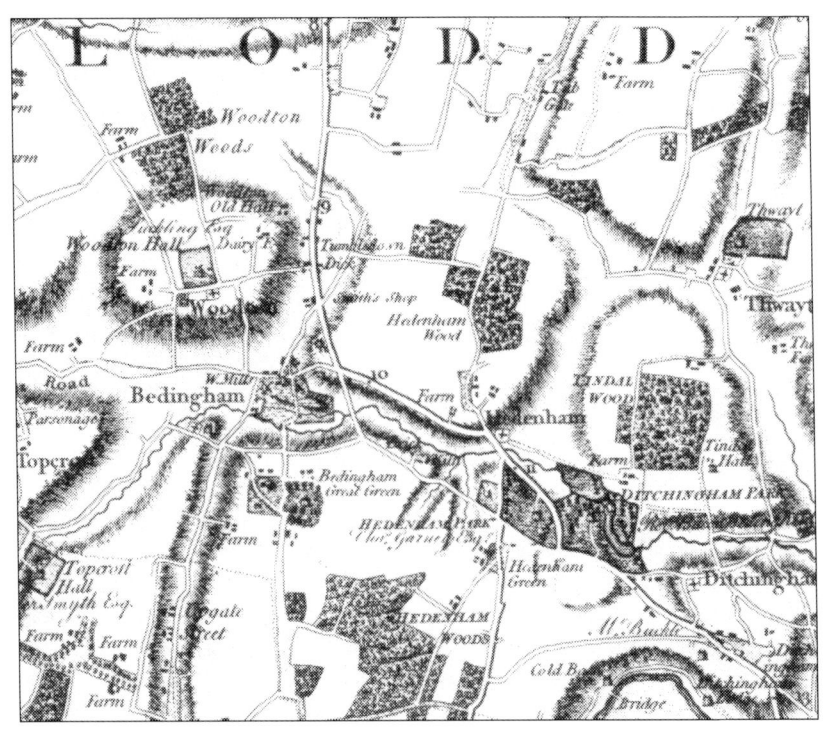

Faden's map of 1797 shows clearly the regional contrasts within the county. This extract shows the heavy clays: an area criss-crossed by winding roads where enclosure of the open fields had already taken place and blocks of woodland survived.

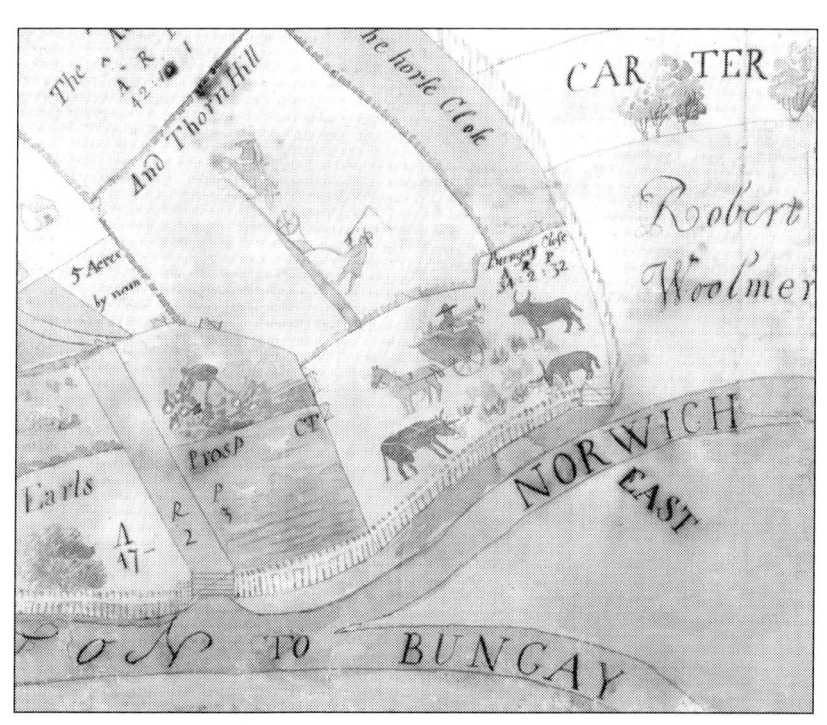

Part of a map of Park Farm, Earsham, drawn probably in the early eighteenth century and showing turnips being fed to cattle in the fields.

pollarded trees. As well as an abundance of hedge timber, there were also many woods, both large and small. The heavy soils were more suitable for cattle fattening and dairying than cereals and so small enclosed fields were required. Probate inventories suggest that most farmers kept milking cattle, fattened bullocks and possessed cheese-making equipment. Few farms had more than a quarter of their land under the plough at any one time.

Although most of the open fields would have disappeared in the hundred years before 1670, many commons remained. Their enclosure was more complicated than that of the strip fields which could be divided between the individual proprietors. A large number of commoners had rights across the common and even the landless poor could collect fuel and dig clay. All these demands had to be recognised in any reallocation of the land.

The clayland countryside of 1670 was dominated by small hedged fields, the majority of which would have been in pasture, interspersed with areas

CHANGING AGRICULTURE

of common. The cropped fields contained wheat and some barley as well as the peas and vetches suited to heavy land, either used as a fodder crop or ploughed in as green manure. Clover and turnips were also beginning to be grown in small quantities, again as fodder crops rather than incorporated into formal rotations. Between the fields were areas of woodland and commons, often with the farms arranged around them. Many of the farms were small and owner-occupied, although landowners such as the Duke of Norfolk held estates in the area. The farmsteads themselves consisted of timber-framed and thatched houses and barns, possibly with a separate stable and small cowhouse.

Shelfanger Hall Farm in south Norfolk was enclosed by 1720 when this plan was drawn. However, strips still survived in the Lammas Meadows by the river (shown as an insert in the bottom left hand corner).

THE ORIGINS OF 'NORFOLK AGRICULTURE'

The north-east heaths and loams

In contrast to the clays to the west and south, this was an area dominated by arable farming. Wheat, rye and particularly barley were grown. As we move north into the less fertile soils we find large areas of heaths and commons taking up about 17% of the northern heathlands where cattle and sheep were grazed. Here enclosure was advancing across the open fields in a slow and piecemeal manner, so that in 1670 a substantial amount of

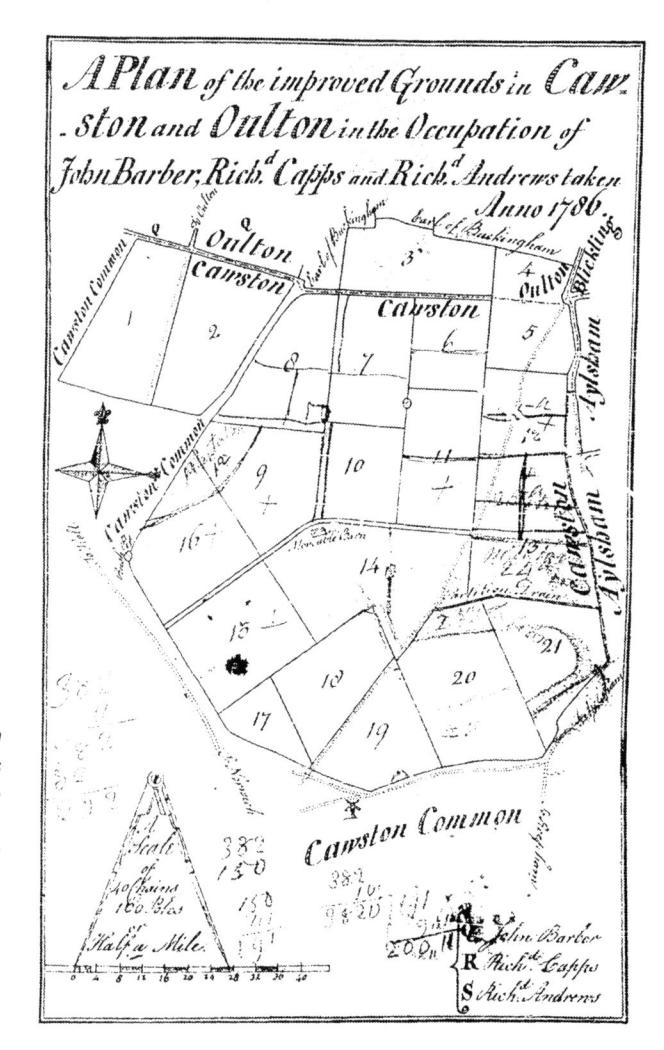

By 1786, when this map was drawn for the Heydon estate, the deer park at Docking in Cawston had been enclosed into regular fields.

17

CHANGING AGRICULTURE

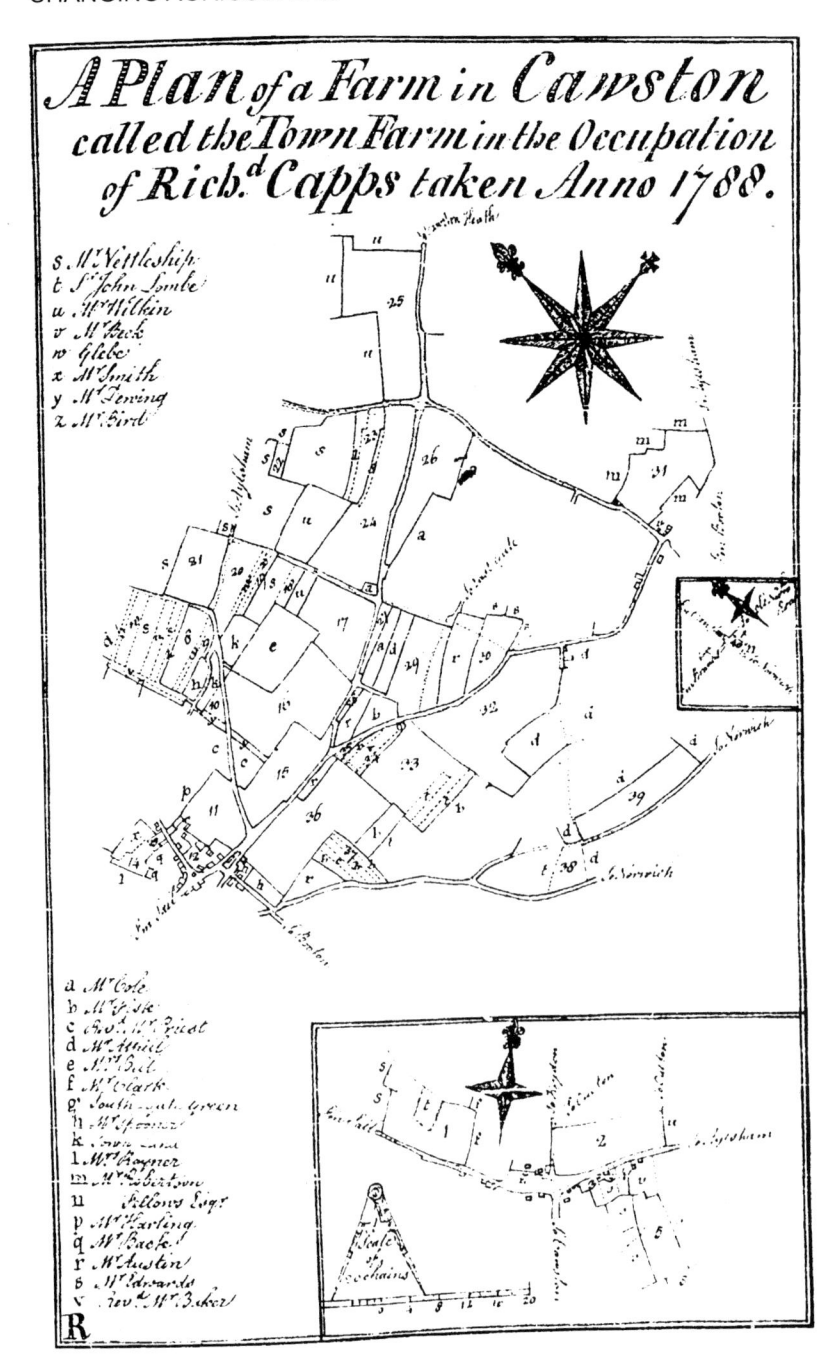

A Plan of a Farm in Cawston called the Town Farm in the Occupation of Rich.d Capps taken Anno 1788.

References

To the Farm in the Occupation of Rich.ᵈ Capps.

		A. R. P.			A. R. P.
†	1. Sydenham's Meadow	7.3.7	22.		-.-.2.21
†	2. Dove House Meadow	7.-.15	23.		2.1.7
†	3. Walnut Tree Piece	-.2.23	24. New Close		11.1.28
†	4.	-.2.5	25. Coxes		12.1.38
†	5. Four Acres	3.2.30	26. Eight Acres		9.-.10
	6. } South Gate Field	-.2.37	27. New Piece		-.-.3.24
	7.	-.1.22	28. Ditto		-.1.15
	8. } Four Acre Field	2.1.36	29. Field Well		6.-.38
	9.	-.3.4	30. Three Acres		3.2.9
	10. Three corner'd Piece	-.1.31	31. East Gate eight acres		8.-.2
	11. Home Pasture	5.-.30	32. Fifteen Acres		15.3.13
	12. Homestall	2.1.8	33. Clamp pit Close		8.-.6
†	13. Cottage Yard	-.1.15	34. Clamp pit Hole Piece		-.1.0
†	14.	-.3.6	35.		-.-.21
	15. Six Acre Piece	7.-.37	36. Pound Close		12.2.32
	16. Brick kiln Piece	10.3.34	37.		-.1.35
	17. Six Acres	6.3.3	38.		1.3.34
	18.	-.3.20	39. Three Acres		3.1.10
	19. } Aylsham Field	-.1.2	40. Brandiston Field		-.-.39
	20.	3.1.7			165.1.18
	21. New Four Acres	4.3.12			

+ Those pieces mark'd thus are taken from this farm except Saver Nᵒ 31 & ... exchanged for other Lands & 3 Rood ...
of William ... to it — makes 149.2.37

Town Farm, Cawston, contained a mixture of enclosed and open land. (Areas are expressed in acres, roods and perches – one acre consisted of 4 roods, and one rood consisted of 40 perches.)

CHANGING AGRICULTURE

Old Hall Farm, South Burlingham, is an example of a manorial farm in the good Norfolk loams. The fine seventeenth-century house exemplifies the high status of the farmer and the large eighteenth-century brick barn indicates the importance of grain within the economy.

open field still survived. Some survived into the eighteenth century as is shown on Corbridge's map of Wolterton drawn in 1732. Low land prices on these heathy soils allowed landowners to accumulate sizable estates and the Hobarts of Blickling, the Harbords of Gunton, the Earles (later Bulwer Longs) of Heydon and the Walpoles at Wolterton were all consolidating their estates in this area. This was very much an 'industrial' countryside with wealth derived both from the manufacture of cloth and commerce and from agriculture. In 1722 Daniel Defoe described the 'several good market towns, and innumerable villages, all diligently applying to the woollen manufacture, and the country is exceeding fruitful and fertile as well in corn as in pastures'. It is the decline of these industries over the next hundred

THE ORIGINS OF 'NORFOLK AGRICULTURE'

years and the resulting surplus of labour which, as we shall see, is one of the reasons why an intensive agriculture was able to develop in Norfolk. Agricultural practice was already changing in the 1720s. Defoe stressed the importance of pasture as well as arable and while barley was always the most valuable crop, inventory evidence shows that cattle were certainly a valuable part of the business of many farmers. John Bowells of Cawston who died in 1710 ran a malting business, but also owned three horses, five cows, three heifers, two calves and two hogs. Not only did he grow barley, but rye, peas, oats and a 'parcel of turnips' for feed. Richard Hill of Saxthorpe was also growing turnips for feed in 1716, and in the 1720s Erasmus Earle of Heydon was comparing the prices gained for his grass and turnip-fed cattle. It is clear that the new crops were already finding favour in this region of the county.

To the south, the loamy soils were far more fertile and the region densely settled. By the thirteenth century, its agricultural system was already one of the most advanced in the country, with crops carefully weeded and animals stall-fed on fodder crops. As a result grain yields were remarkably high. The richness of the soils, coupled with the weakness of the manorial system, helped prevent the growth of large estates and allowed freeholders to establish their rights. Farming here was mixed, with cattle fattening and dairying more significant than sheep. The livestock was grazed in the summer months on the rich grazing marshes of Broadland which, according to Defoe, grew up to 'the fattest, though not the largest in England'. A highly productive system, providing a good living for its well-to-do owner-occupiers, did not encourage enclosure and so this is one of the areas of the county where open fields lasted longest, in many parishes into the nineteenth century. In contrast therefore to the claylands, this was an area dominated by arable land farmed in open fields. Large herds of cattle would spend the summer on the marshes and winter on the stubbles, or indoors in sheds. The wealth of the farmers can still be seen in their fine brick yeoman houses and farm buildings. The survival of early eighteenth century brick barns and stables is an indication of the standard of building at this relatively early date. It was to this area and these independent farmers that the agricultural writer William Marshall, writing in 1787, traced the origins of 'Norfolk Agriculture'. 'In East Norfolk alone we are to look for that regular and long established system of practice which has raised deservedly, the name of Norfolk husbandmen, and which, in a principal part of this district, remains unadulterated to the present time.' Research on probate inventories by Mark Overton supports Marshall's views. New crops such as turnips were certainly being grown in the north-east by the late seventeenth century. However, they were only grown in small quantities, primarily as a fodder crop in existing mainly pastoral systems, rather than specifically as part of an innovative rotation. Only gradually were they taken up in the arable, sheep-corn areas of lighter soils to the west. Over Norfolk as a whole, clover only made up 3% and turnips 8% of the cropped acreage as late as the mid-eighteenth century.

North-west Norfolk

This area of light free-draining soils was, by the nineteenth century, claiming to be the home of the Norfolk 'agricultural revolution'. However in the seventeenth century it was sparsely populated and supported a cereal and sheep system, dominated by the great estates whose owners had been able to accumulate the relatively cheap land on these poorer soils. Here Chief Justice Coke built up his family fortunes around Holkham, and Robert Walpole invested his profits as prime minister around the family seat at Houghton, next door to his political rival Charles, Viscount Townshend at Raynham. All these names were to feature in the agricultural history books. The long established Le Strange family owned land around Hunstanton, whilst many less well-known families owned sizable estates across the region. Here too large areas of open field remained until the end of the eighteenth century, but for a different reason to that on the fertile loams. The result was a landscape of huge empty open spaces dominated by common and open field, few roads, sparse settlement and expanding areas of park being created by the gentry.

The foldcourse system of East Anglia, particularly of north-west Norfolk and Breckland, distinguished the region from the other chalkland sheep areas of the south of England. In both regions the fertility of the arable depended on manuring by the sheep, and in East Anglia this was highly regularised. The sheep flocks were mainly owned by the lords of the manor or their lessees. In the growing season, the sheep would be confined to the commons and heaths, but for the rest of the year they would be hurdled onto the common fields. These rights of 'foldcourse' went with individual manors and often became subjects of disputes between the manorial lord and his tenants in the open field. The system worked well while the normal time of sowing was the spring and when no winter crops were grown. It provided the tenant with a 'free' source of manure and while it suited the landlords well, it was unlikely to change.

THE ORIGINS OF 'NORFOLK AGRICULTURE'

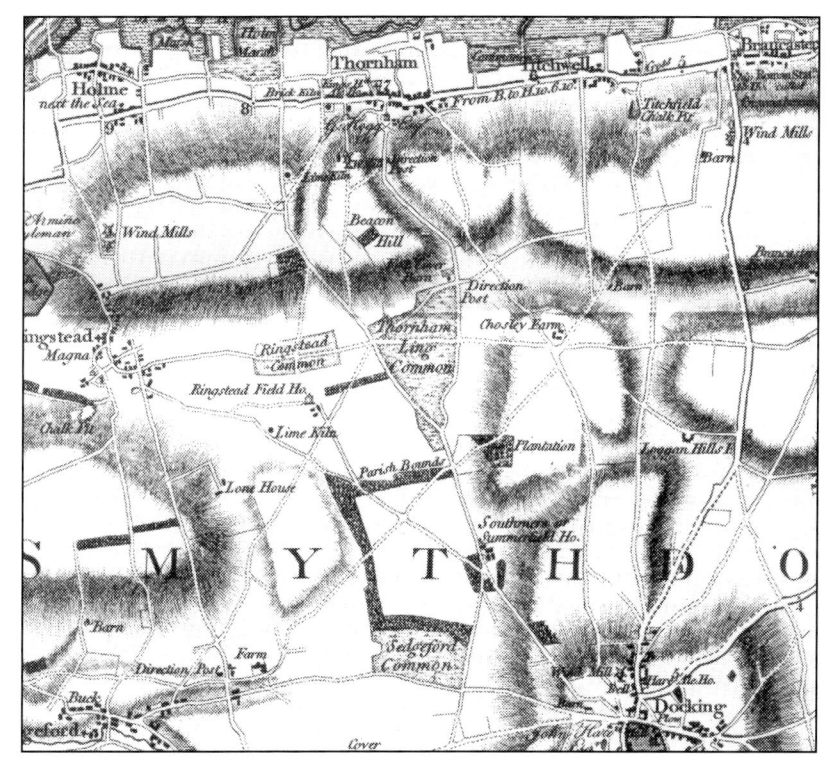

This extract from Faden's map depicts the wide open countryside of much of northwest Norfolk. Population is sparse, much of the land is still unenclosed (as indicated by the dashed lines marking road edges) and commons remain. There is little woodland and the roads themselves are comparatively straight.

From the early eighteenth century the number of disputes increased. Tenants wished to try new crops such as turnips and keep their own animals on them in the fields. Winter-sown crops were also being introduced as well as artificial grasses such as clover. The records of Red Barn Farm, Snettisham, which was owned by Kings Lynn Borough, show that in the seventeenth century the farm consisted of 49 acres of enclosed pastureland, 104 acres of 'furfield' (outfield?) land and 240 acres of open field with a foldcourse for 800 sheep, but in 1694 there was a dispute over 'that part of it now sown in clover'. It was claimed that for the last forty years it had been used as shackage (another word for sheep feed after harvest) and so must return to that use. Similarly records in the Norfolk Record Office show that in 1673, in the parish of Anmer, Richard Potter was forced to open up

23

CHANGING AGRICULTURE

a piece of the open field he had enclosed after it was found 'by the examination of ancient witnesses as also the reading of a certain description of witnesses made in the High Court of Chancery . . . that the said close is a shack close and so has been anciently used'.

As the eighteenth century wore on, it became clear to landlords that it would be more profitable to get rid of their sheep and the shepherds they had to employ, and enclose the open fields that could then be let at a much higher rent to the new breed of capitalist farmers who would then farm intensively, marl and manure the land and introduce new crop rotations to produce crops adequate to support the higher rents. It was the high grain prices and accompanying rent rises of the Napoleonic wars that finally brought the foldcourse system to an end.

Detailed records surviving from the Le Strange estate at Hunstanton from the seventeenth century as well as some references from the Walpole and Raynham estates allow us to reconstruct the farming of this area at the beginning of our period. When Charles Townshend inherited his estates at Raynham in 1687, they were still mostly farmed on an open-field system with the landowner owning huge flocks of sheep which he grazed across the fields and commons at Coxford, Creake, Stiffkey and the 'Great Ground' and 'Granoe Hill' at Rudham in 1699. His lambs at Creake were sold for £198, so they must have come from a sizable flock. During his early years Townshend was responsible for enclosing much of his land and there are frequent references in the documents to ditching, hedging, marling and manuring. The letters from his agent, the Reverend William Priestland, describe getting the hedges of Little Raynham in order and the digging of marl pits. Once the land was enclosed, tenants were granted leases requiring them to farm according to improved methods. Leases included mention of turnips in rotation with other crops. In 1698 Robert Tickler of East Rudham committed himself to divide Thorn Close into three enclosures and sow one of them with turnips every year. New farms were laid out and new buildings erected. We read that Robert Tickler was 'very pressing for the rebuilding of his barn'. Meanwhile estate records show that Townshend was giving up his flocks and relying for his income far more on the rents of his newly enclosed farms. A similar picture emerges from the records of the Walpole estates and it is clear that a trend towards the letting of enclosed farms under leases was spreading across Norfolk estates, both great and small, from the late seventeenth century. The little evidence that survives, mostly for the better-run estates, shows a consistency that suggests that there was a general move towards leases with husbandry clauses encouraging the growing of turnips from the 1680s. The fact that many of these well documented estates are in the west, while less evidence survives elsewhere, must not mislead us into assuming that this was the only region where such changes were taking place. It may simply be that the records for small estates and owner-occupier farms do not exist.

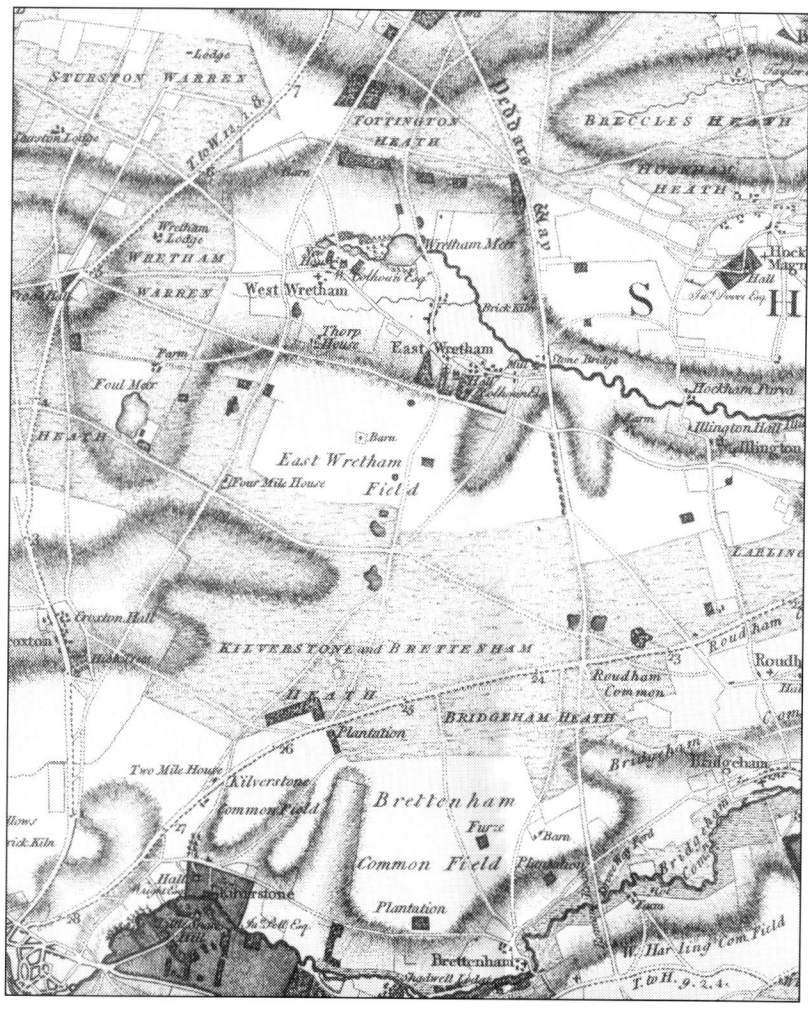

The Breckland landscape is even emptier than the north-west. Villages such as West Wretham are clustered along the valleys with their open fields behind (East Wretham Field), cut out of areas of warren (Wretham Warren) and heath (Kilverstone and Brettenham Heath).

CHANGING AGRICULTURE

Barley was the dominant crop on the better soils with rye restricted to the poorer areas. Wheat and peas were extensively grown and clover was already a common crop. Land was usually fallowed every four or five years.

Breckland

Breckland is one of the most distinctive agricultural regions of East Anglia and lies in the extreme west extending into Suffolk and Cambridgeshire. It consists of very sandy soils on porous chalk and so is the least fertile region. It also suffers from a dry and often cold climate in which frost can occur in any month of the year. In the seventeenth century much of it was open heath and rabbit warren. The term 'breck' means an outfield which was only cultivated from time to time and this practice was typical of the area. The arable fields were open with extensive rights of foldcourse across them which were of great economic importance to their owners. The Iveagh records show that in 1624 Robert Lord made the not inconsiderable sum of £522 from his Elveden (Suffolk) foldcourse in skins, pelts and wool as well as hogetts, lambs and crones (old ewes). Slightly better soils are found in the valleys and here around villages some enclosure had taken place and crops grown. As in north-west Norfolk some farms were already huge, particularly where they occupied almost an entire village as at Cranwich, near Mundford. Again as in north-west Norfolk, much was owned by large landowners who, as the eighteenth century progressed, hoped to make profits similar to those being made by their eastern neighbours; as we shall see, they made costly, and in the long term mostly fruitless, efforts to make the windblown sands produce high yielding cereal crops.

The Fens

Another area awaiting the attention of 'improvers' in the seventeenth century was the fens. As a result of the General Drainage Act of 1600, it was possible for large landowners to overrule local proprietors and suppress any common rights which obstructed drainage schemes. However, there were no grand projects in Norfolk to compare with Vermuyden's work in Cambridgeshire and much of the peat fen remained a wet and empty landscape, utilised by wild fowlers and reed cutters until the nineteenth century. To the north of the peat area which had accumulated behind the silted-up mouths of the Ouse and Nene, the silt fen edge was fringed by wealthy

THE ORIGINS OF 'NORFOLK AGRICULTURE'

The fens in the north-west of the county formed a distinctive area in the 1790s. Villages such as Walpole and Terrington stretched out along the drove roads leading into Marshland Smeeth, which was used for summer grazing.

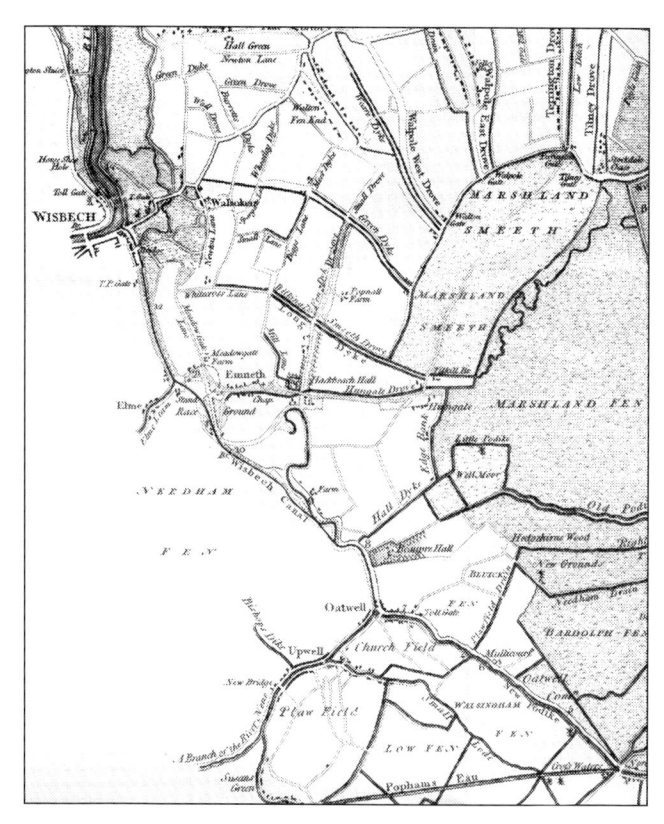

linear villages such as Feltwell, Outwell and Upwell. Their farms had gradually been nibbling at the surrounding fen and creating a patchwork of tiny, often long and sinuous fields enclosed by ditches and banks. Drove-ways through them allowed access to the gradually diminishing area of grazing marsh of Marshland Smeeth behind. The newly created fields supported a mixed farming, with a great emphasis on grazing.

Progress of change 1670–1720

The farming landscape of 1670 was therefore one of great variety, ranging from the open wastes of Breckland so dreaded by travellers from the south and west, to the mainly enclosed and wooded pasturelands of the clays. Open fields survived both in the fertile arable and cattle areas of the east and in the light sheep county of the west where the foldcourse was still fun-

CHANGING AGRICULTURE

damental to the farming system. Some of these regional differences were becoming blurred as enclosure, particularly on the estates of the northwest, began to pick up by 1720. New crops such as clover and turnips also began to find their way from the owner-occupiers of the east to the tenant farmers of the west, sometimes with the encouragement of their landlords. The period after the restoration of the monarchy in 1660 saw a development of interest in new farming methods. An increasingly literate yeoman and gentry class was reading the many new books on farming, some of them translated from the Dutch (the Dutch being considered masters of intensive agriculture). Both farmers and landowners from Norfolk had many contacts in Holland. Charles Townshend, for instance, led a government delegation to The Hague, spending time there during the lengthy negotiations culminating in the Treaty of Utrecht (1713). It is this combination of a prosperous, intelligent, educated gentry and farming elite, with the high profile activities of some eminent landowners, which will be considered in the chapters which follow.

THE FORCES OF CHANGE 1720–1840

Landlords and 'improvement'

The period 1720 to 1875 is one in which the landed classes dominated rural Britain. By the end of this period over half of Norfolk was owned by 223 landowners with more than one thousand acres each. These estates varied enormously in size with the smaller ones often consisting of little more than a home farm and a couple of tenants while the largest, that of Lord Leicester at Holkham, covered 43,000 acres and 70 farms. His two near neighbours, the Townshends at Raynham and the Walpoles at Houghton, owned between 15,000 and 18,000 acres each. Below these giants were a further eight estates of over 10,000 acres, and 25 of between 5,000 and 10,000 acres, most of which were also in the north-west of the county. Records of only a minority of these estates survive, and are likely to represent only the best-run of them. However, these show a remarkably similar type and timing of development, which suggests that the activities on the undocumented ones may well have been similar. Because we are only able to follow events in a handful of estates offices, such as those at Holkham and Raynham, it would be a mistake to assume that these were somehow unique and that similar interest was not being taken by other less well recorded landlords.

By the middle of the eighteenth century commentators were clear as to what constituted the improved farming of Norfolk, and claimed that the changes had been gaining ground for about a hundred years. The characteristics of this farming system described by a writer to the *Gentleman's*

Magazine in 1752 are very much the same as those listed by Arthur Young in 1771. Young saw the distinctive elements of Norfolk farming as:

- Enclosure with or without the assistance of Parliament
- The proper rotation of crops
- The use of marl or clay
- The cultivation of turnips, hand-hoed
- The cultivation of clover and artificial grasses
- Long leases
- Large farms

This list will provide a suitable starting point for our consideration of the county's agriculture between 1720 and 1840 and the items can be divided rather crudely into those which were the responsibility of the landlord and those which fell to the tenant. Those which involved the landlord will be discussed in this section, and those which involved the farmer will be considered in the next.

The landlord–tenant system under which most agriculture operated was seen by contemporaries as ideal. The tenant's capital was used for the working and the stocking of the land, while the landlord was responsible for fixed capital investment such as buildings and fencing, and it was claimed that this resulted in an efficient and highly capitalised agricultural industry. Following Young's definition, the landlord would be responsible for enclosure, leases and the size of farms, whilst the actual method of farming, the crops grown and in some cases the marling would be the responsibility of the farmer. The belief in the benefit of the dual roles is underlined by the toast, 'A good understanding between landlord and tenant' which was drunk at the Holkham audit dinner.

Although much stress was laid on the mutual advantage of the landlord-tenant system there were of course strains, particularly when it came to setting rents. In times of agricultural prosperity when farming seemed a profitable venture, tenants were easy to come by and high rents could be demanded. The landlord could pass on the responsibility for maintaining soil quality by marling and draining to his tenant and enforce strict husbandry regimes to ensure or even improve on levels of fertility, through the terms of the leases. As soon as prices fell, then it was tenants who had the upper hand, because no landowner wanted to leave a farm tenantless and for land to become derelict. Thomas De Grey, acting as agent for his brother on his breckland estates around Merton, wrote to his brother in London during a run of bad years in the 1780s, 'Norfolk farmers have long since ceased to be humble dependents; we must therefore bear with their language which is to be treated on equal terms. If you wish to be master of your estate, you must reject Mr Smith; if you wish for ease and less profit you must accept him on the best terms that can be made', and again, 'The time has been when landlords dictated to their tenants, but that has passed long ago . . . I know not how to advise – in one of the Norwich papers last week only, there were 27 sales of the various stock of farmers'.

'Improvements' were carried out both to allow high rents to be charged and to attract tenants with capital who could, through restrictive covenants, be compelled to farm in such a way as to keep the land in good heart. This in turn kept land values (and rents) up. The often sycophantic descriptions of the achievements of some land owners need to be put in context. The main reason for the 'improvements' they carried out on their land was to increase the value of their assets and to raise rents.

But it would be wrong to be entirely cynical about the motives of all landowners. There were those who believed that it was their moral obligation to increase the output of their estates, whilst at the same time providing employment for the poor. The Scottish landowner, Lord Belhaven, one of the earliest advocates of improvement, wrote in 1699 that the increasing of food production was a patriotic duty: 'Husbandry enlarges a country and makes it as if ye have conquered other countries adjacent thereto. And I am sure a conquest by the spade and plough is both more just and of longer continuance than what is got by sword and bow.' One of the earliest literary references to the value of increasing agricultural output is to be found in *Gulliver's Travels,* first published in 1726 and written by the political satirist Jonathan Swift: 'He gave it as his opinion that whoever can make two ears of corn or two blades of grass grow upon a spot of land where one grew before, would deserve better of mankind and do more essential service for his country than the whole race of politicians put together.' The economist Adam Smith regarded the improvement of agriculture as the most important employment of capital. There are many more similar examples in the literature of the time, and their sentiment is reflected in a letter of Thomas De Grey in which he says of reclamation work on the sandy heaths, 'the great expense would but ill answer, unless there were real satisfaction in employing labourers and bringing forth a ragged dirty parish into a neatness of cultivation'. The degree to which a landowner took these duties seriously was evident for all to see. A landscape enclosed into large rectangular fields divided by hawthorn hedges, dotted with new sets of farm buildings and a few (often very few) cottages, drew to the owner the admiration of his neighbours and friends who visited, walked and rode his estate with him during the shooting and hunting season.

Landlords and enclosure

We have seen that on the Norfolk estates for which we have evidence, landlords were already beginning to enclose fields and create farms which they let on leases by 1720, but this trend certainly increased thereafter, and by mid-century the 'improvement' of estates was a fashionable gentlemanly pursuit as well as a profitable pastime. How significant all this activity was on the county as a whole, or to long-term farming change, is something we shall need to consider.

The landlords whose names are inextricably linked with the promotion of the 'Norfolk system' of agriculture are Charles second Viscount Townshend, who had probably acquired the nickname 'Turnip' within his own lifetime, and Thomas William Coke ('Coke of Holkham'). Although these two names are often linked, their lives were in fact separated by a generation. Townshend, who owned his estates from 1687 until his death in 1738, was one of the first generation of landlord improvers. For much of this time he combined estate management with a successful political career in which he rose to being one of the two Secretaries of State, his neighbour Robert Walpole being the other. Presumably it was his high public profile that gave such publicity to the changes he was making on his estate, but as we have seen his activity as an encloser and encourager of new farming methods was no more than was being carried out by many other landowners, both great and small. Turnips were certainly being grown elsewhere in the county, on both owner-occupier and estate farms, although not in large quantities, and it is likely that rather than being an innovator, Townshend and his fellow landowners were simply encouraging best practice in an effort to raise yields without exhausting the soils and so increase rents. Townshend must have been one amongst a sizable group of farmers and landowners who were innovating in their enclosed fields.

Thomas William Coke held Holkham from 1776 until 1842 and was created Earl of Leicester of Holkham in 1834. It is he who is remembered for his

THE FORCES OF CHANGE

improvements, but much had been achieved by his predecessor, another Thomas Coke, who held the estate from 1718 to 1758, and whose activities at Holkham are reminiscent of those at Raynham. The completeness of the documents at Holkham make it possible to assess the impact of the changes. It is clear that Thomas Coke's improvements were indeed very profitable as rents went up by 44% over the forty years of his ownership. Dr Parker has argued that as there was very little rise in prices during this time this increase can only have been made possible by an increase in productivity resulting from more intensive farming.

One way of doing this was through enclosure of the strip fields. The account books are full of sums of money for fencing. In 1736, for example, Carr of Massingham was allowed £18 for six cords of battens for his new enclosures (possibly for building the gates) and by 1752 there was only one unenclosed field left on his farm. The estate promised William Lee, tenant of Longlands Farm, Holkham, that it would 'inclose and divide into proper Inclosures as was verbally agreed, all those lands at present uninclosed in the space of 2 years'. The speed of enclosure varied from farm to farm and would have depended, amongst other things, on the number of freehold intermixed strips that had to be bought out. By 1779 when a series of estate maps was produced, very few strips remained. The profitability of this reorganisation for the landlord is clear. Rents could as much as double as a result of enclosure.

As well as enclosing strips, there were the open sheep walks which had only intermittently been cultivated. Here enclosure had to accompanied by other methods of soil improvement if these light lands were to be successfully cultivated. Marling involved the spreading of a chalky, clayey subsoil on the fields. The clay would help make light soils more moisture retentive, while the chalk would reduce acidity. It is clear that Thomas Coke was financing the spreading of quantities of up to thirty loads per acre on these light-soil fields. The growing of turnips to feed large flocks of manure-producing sheep was also an important element in this land reclamation. The hoeing of turnips at the Holkham home farm was one of the most expensive tasks on the farm, costing £25 15s 6d in 1730. The Home farm accounts show that marling cost twice as much at £52, and taking up a fifth of farm expenditure on labour that year.

By the time Thomas William Coke inherited the estate in 1776, little open field or sheep walk remained to enclose, but there were still open commons. These often consisted of the worst land in the parish, but by 1816

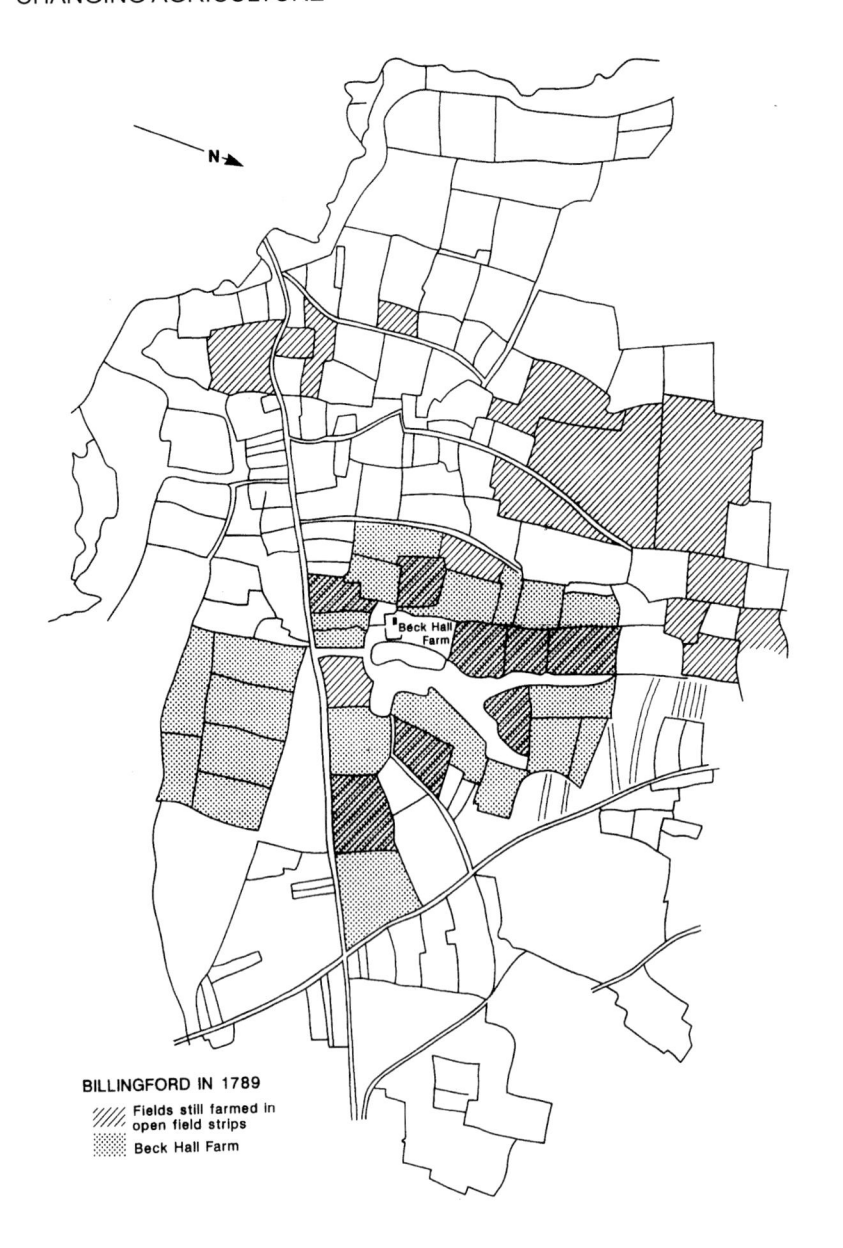

CHANGING AGRICULTURE

BILLINGFORD IN 1789
▨ Fields still farmed in open field strips
▦ Beck Hall Farm

In 1789 much of the parish of Bllingford in mid-Norfolk was still farmed in strips. Beck Hall farm had both strips and enclosed field scattered across the parish and these are shaded on the plan.

THE FORCES OF CHANGE

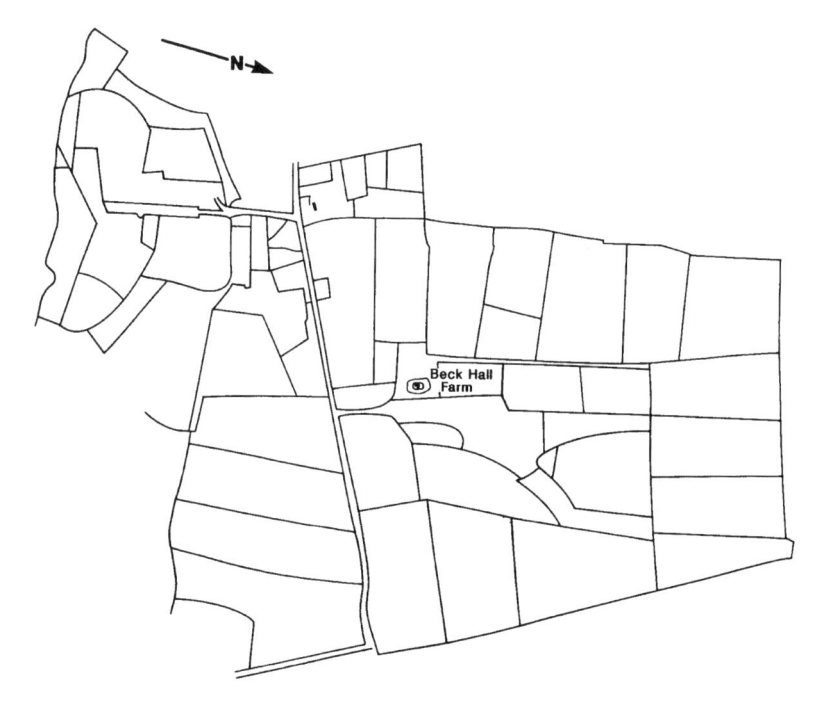

By 1830 the gradual enclosure of Beck Hall and the consolidation of its fields around the house had been completed; most of the old field boundaries had gone.

they too had nearly all been brought into cultivation and by then the rents on the estate had doubled. This is not as spectacular as it might seem as it was a period of rapid inflation, particularly of grain prices during the wars against the French between 1796 and 1815. Because of the profitability of agriculture and the high demand for farms, Coke was able to pass on the expense of marling and draining to his tenants. Before the introduction of clay pipe drains in the 1840s, the drains would have been filled with furze and bush, and their regular digging was not as important an element of improvement on the light soils as on the heavier lands in the county, but it could still be vital to the increasing of yields. The years immediately following the end of the Napoleonic wars in 1815 saw the fall of grain prices and a loss of confidence in farming, and it could be difficult for the landlord to force tenants to undertake these, often expensive, tasks. John Hastings of Longham, where the heath had only just been enclosed, could not afford the work involved in bringing the land into cultivation and so was promised support from his landlord.

CHANGING AGRICULTURE

It would be a mistake to believe that it was only on the largest of the Norfolk estates that efforts were being made to bring more land into cultivation. The profits to be made during the Napoleonic wars encouraged most owners to maximise their farm income. The enclosure of the Hare estate around Stow Bardolph was completed by 1801, but according to John Wiggins, the new agent appointed in 1811, there were 'many bad farmers', and buildings and land were neglected. Between 1811 and 1821 much of the 'upland' on the edge of the fens was fenced, marled, drained and supplied with new buildings. Bricks from a new brick kiln were used to replace the former 'ill-constructed buildings'. New cattle yards helped keep the stock warm and conserve manure. Yearly tenancies were replaced by 21 year leases which required the tenant to plant and maintain hedges, put in new drains, clay (or marl) the fields, and maintain and repair ditches and buildings. Twice-yearly visits by the agents made sure that these terms were being adhered to. As a result rents rose from £7,000 in 1809 to £11,131 in 1821.

Although only a limited amount of enclosure remained to be done at Holkham, Coke did not stint his expenditure on improvements, much of it on the new farm buildings which form such an important element in the present-day landscape. The providing of sound and well-arranged farm buildings was seen as one of the duties of a landlord. The cattle needed for producing manure could not be kept unless suitable housing in sheltered yards was available. The Holkham estate farm at Waterden with its huge L-plan barn and extensive cattle yards and stables, was described by Arthur Young in 1784: 'Every convenience to be imagined is thought of, and the offices so perfectly well arranged as to answer the great object to prevent waste and save labour'. Francis Blaikie, agent to the Holkham estate from 1816, wrote of Waterden that it had 'perhaps the finest set of farm premises in Great Britain'. Between 1790 and 1820 about 30 of the 70 Holkham farms went through major rebuilding with 15 being completely replaced at costs varying from £1,500 to £3,500. These involved the creation of care-

Waterden farm: the basic plan survives from Arthur Young's time, though further gable-ended sheds were added in the 1870s to divide up the cattle yard.

THE FORCES OF CHANGE

fully planned farmyards with huge barns and shelter sheds for housing cattle through the winter. Owners of some smaller estates could not afford such grandiose rebuildings and here, as on the Brampton Gurdon estate around Letton Hall on the Norfolk claylands, earlier timber-framed buildings survive. Absentee landlords such as the Duke of Norfolk took little interest in their farms and the result was that by the 1860s most of the buildings needed completely replacing. A report of 1861 pointed out that the policy of previous dukes which had involved simply taking rents and doing nothing meant that they had been able to keep most of the rent collected, but rents could not rise.

Long leases were also seen as fundamental to the Norfolk system by Young, and one innovation for which Coke claimed much credit was 'the Holkham lease' which became the standard form of agreement across Norfolk and beyond by the 1840s. In fact, leases had been normal practice at Holkham as elsewhere long before Coke. As a result of advice given by the London land agent Nathaniel Kent in the 1790s, and then under the Holkham agent Francis Blaikie after 1816, they became far more formalised. They were usually for 21 years and laid down a specific crop rotation to be followed which would take six years to complete. In the first year

The granting of a lease, as shown on the Holkham monument (1845).

CHANGING AGRICULTURE

The sheep shearings, as shown on the Holkham monument (1845).

turnips or vetches were grown which could be lifted for cattle or fed off by sheep during the winter. Spring barley undersown with grass seed was then planted. After the barley harvest, the land would be left in grass for two years before being ploughed up in the autumn and sown with wheat; after this the cycle would begin again. Later leases became more detailed, specifying the hoeing of the turnips twice (to clean the land of weeds) and also the amount of grass seed to be sown. In 1816 Francis Blaikie wrote a series of leases giving different types of rotations suited to different soils and these survive in a bound volume at Holkham. On the best soil, land did not need to be left in grass more than one year, thus allowing for a 'four-course' rotation, and this 'Norfolk four-course' was the one universally taken up by the mid-nineteenth century.

While it is clear that many of the improvements for which Coke was famous had been going on for some time before he inherited, there is one for which he could justly claim fame and that is the Holkham sheep shearing. These private agricultural shows, similar to those run by his friend the Duke of Bedford at Woburn, started as local gatherings of his tenants soon after Coke inherited in 1776, and grew to be fashionable international three-day events promoting improved breeds of sheep and innovative agricultural

THE FORCES OF CHANGE

practices generally. The last meeting was held in 1821, at a time when agriculture was in depression after the end of the war against France.

It was probably a high-profile and controversial political career, helped by the publicity he managed to attract at the sheep shearings, that pushed Coke's agricultural activities into the limelight. His survival until 1842 ensured that he would be seen as 'the grand old man of Norfolk agriculture', and after his death a monument to his achievements was erected in Holkham Park.

Whilst we know most about these high-profile, publicity-conscious landlords, there is no doubt that much was going on elsewhere. This is certainly true of enclosure where most activity, both before and after 1660, was carried out by agreement and exchanges between owners, or single-handedly by a landowner who owned an entire parish. Much of this happened with little or no record. Over large areas of the claylands, most enclosure had taken place by 1700, with only residual greens and commons remaining to be enclosed by parliamentary acts in the early nineteenth century. The north-east heathlands differed from those further west in that there were few foldcourses to be broken up and so enclosure was easier. Gradually land was enclosed during the eighteenth century, with a whole variety of types of holdings surviving on the Heydon estate into the 1790s. At one extreme, several of the farms in Corpusty and some in Cawston were still held in scattered strips, some with shackage rights across them. Other farms incorporated newly enclosed common, whilst Docking Farm (Cawston) was a new creation of regular fields within an old deer park. Because the Heydon-estate farms were mostly enclosed without acts of Parliament, the only evidence for the reorganisation of fields is in the allowances against rent which were granted to tenants for the cost of between 500 and 4,500 (and in one case 20,000) whitethorns for hedging. Parliamentary acts were confined to areas of heath, such as Cawston Common where a major programme of improvement took place after its enclosure in 1801. Firstly the heath was ploughed and fir faggots were laid in drains to provide drainage. The records show 'mudd' and 'faggots' being delivered weekly. In May the land was ploughed for a second time and seeds were sown. During the winter more ditching was undertaken. Several new farms were created on the heath, and during the first year of the lease they were let rent-free while the work of marling, draining and fencing was being started. The land was thought good enough to support a six-course rotation with three years being reserved for grass. But the reclamation was not a long-term success; much land reverted to heath when cereal prices fell after the

CHANGING AGRICULTURE

Napoleonic wars. The name 'Botany Bay Farm' given to one of the new creations suggests that one of its tenants thought life there little better than in the Australian penal colony of that name.

Areas where parliamentary enclosure was most important were the contrasting areas of the rich loams of the north-east and the poor sands of Breckland, where most enclosure took place after 1790.

Attempts at development had begun in Breckland by the 1770s and indeed one of the earliest examples of the use of the word 'improvement' in Norfolk occurs in an early eighteenth century letter from the owner of a Buckenham farm to the Essex aristocrat and owner of Westwick Hall near North Walsham, Lord Petre, who was trying to consolidate his holdings around Buckenham Tofts. He wrote, 'I have no thought of selling my estate in Stanford. I know that estate is capable of improvement . . . Land is with us a rising commodity and sells every day better than others.' The main obstacle to enclosure of course was the persistence of foldcourse rights, but these were beginning to break down after the 1720s. The foldcourse of the manor of Thetford, for instance, was usually let to the tenant of Thetford Abbey Farm who grazed about 650 ewes and 350 wethers (castrated male lambs). However, by 1724 some of the occupiers of the open field were sowing turnips and preventing access to sheep, and there was a similar dispute in Ashill. The temptation of high rents from wheat lands bounced many a Breckland owner into unrealistic schemes of enclosure and reclamation resulting in the enclosure of 70% of the Norfolk and Suffolk Brecklands during the Napoleonic wars. Much of Ashill was enclosed in 1785 and Thetford Heath in 1804 when Thetford Abbey Farm grew from 800 to 1,896 acres. Lord Petre had owned 277 acres in the open field and this, plus his commoning rights, entitled him to 585 acres of enclosed land.

Breckland, as we have seen, was an area dominated by large estates and so we would expect landlords to have been highly influential in the process of enclosure. The most important of these estates within Norfolk was that of the De Grey family (later Lords Walsingham) of Merton Hall. They were established in their fine Tudor house at Merton by 1600 and gradually added to their estates over the next 250 years. By 1870 they owned 12,120 acres in Norfolk, mainly in the parishes of Merton, Thompson, Tottington, Stanford and Sturston. William De Grey's legal career culminated in his appointment as Chief Justice of Common Pleas in 1771 and he was created Lord Walsingham in 1780. From the 1760s he had little time for his country affairs which were left in the hands of his brother, Thomas, who

THE FORCES OF CHANGE

kept William in close touch with what was going on through a voluminous and lively correspondence. As we would expect, the soils on the estate were mostly poor and this is graphically illustrated in an agreement made with Merton Park's neighbouring farmer: he agreed to 'pay all expenses of clearing sand that may blow from the said piece of ground upon the park'. In preparation for the development of the estate, maps were drawn between 1771 and 1774. They show that a large part of the parishes of Stanford, Sturston and Tottington were open warren, heath, furze and sheep walk. In spite of the fact that these parishes contain some of the poorest soils in Breckland, Thomas wrote to his brother that 'a great part of the land is capable of improvement'. Ralph Cauldwell, the Holkham agent, was less sanguine when he visited the area in 1780: 'When I rode about your estates I observed the best soils towards the extreme parts of your land, and could

A few farmsteads, such as Westmeer, Tottington, were built in the middle of open heathland. The large house, barn with lean-to and stable shown on this map of 1760 suggest a productive farm providing a good living for the farmer. The soil was poor, so the farm must have been an extensive one including not only strips in the open fields but also rights of foldcourse over the land in winter, allowing the farmer to keep very large flocks of sheep.

West-meer Farm House

it be considered to enclose and cultivate for corn and cattle some of these parts and leave the worst for rabbits, it may suit every purpose'. Thomas conceded that it was 'an arduous affair' for 'tenants to make improvements over such a waste country'. Nevertheless, work began in Stanford and Sturston with the landlord planting hedges and building farmhouses and other buildings including barns, stables and granaries. The landlord also paid for the carting of manure to improve the pastures.

Whilst Stanford and Sturston were owned entirely by one landlord and so he could undertake their enclosure and reorganisation and building of new farms without recourse to Parliament, this was not true of Tottington and in this case we are fortunate to have in the records a detailed account of the progress of an enclosure which must be typical of many others in the county. At first De Grey hoped that he would be able to buy out the other, mostly small holders in the parish. One, Mrs Duffield, who owned 43 acres in 33 separate strips, realised how strong her position was and would not agree to the price she was offered: 'she says, (and with some truth) that she can increase the rents, though I cannot, because she can let her little scraps to such tenants of mine with whom they are intermixed and they must give her the full price or suffer by refusal.' Although a deal was eventually negotiated with her, the real problem was four small holders with less than ten acres, and because they refused to sell, a parliamentary act had to be sought. The three commissioners met in 1774, all local men, presumably of local standing, and drawn from the gentry class. Thomas De Grey thought they would prove to be 'friends of the little proprietors'. Of the four small holders, two were described as 'yeomen', one a carpenter and one as 'gentleman'. Only one lived in Tottington although the others were local. It is probable therefore that they all let their land rather than farmed it themselves. Although it cannot be denied that enclosure would cause an upheaval, it was not necessarily the small freeholders who were pushed around or lost out; rather it was their tenants. Peter Dent who had owned nine acres was allotted 24 to make up for his loss of commoning rights, Thomas Levett who had owned five was allotted 12, William Balls who had owned two was allotted 13 and William Neale with less than one now had nine. As De Grey was the largest owner, most of the expense fell on him, as did the work of fencing and reorganising the original three farms he had owned into five new ones, dividing the land and erecting buildings.

Here in Breckland we see the power of the single landlord to 'improve' at its extreme, but when we consider the resulting farming practices we will see that even in this situation and with this autonomy he was not in the long run able to create a sustainable system of arable farming on land entirely unsuited to it.

The fertile loams were the other area of late enclosure. In contrast, the rents here had always been high and there was little incentive to enclose at an early date. There were no dominant landowners which also made change difficult. Over half of the open field in the Flegg Hundreds remained to be enclosed by parliamentary act – in some cases, as at Winterton, not until well into the nineteenth century.

Enclosure of the commons

The enclosure of commons by agreement was much more difficult as many people would have a variety of commoning rights which really needed the force of law to sort out. The commons were the last land waiting to be brought into intensive cultivation and many disappeared during the Napoleonic wars. Three hundred enclosure acts dealt with Norfolk, but many merely involved small areas of common, or tidying up enclosures that had mostly taken place. Most of the acts in the east, away from the loams, involved only common and most dated from the years of the Napoleonic wars.

The result was that by the time of the tithe maps in the 1840s nearly all the open fields and most of the common land, which had been a feature of the Norfolk countryside since the middle ages, had disappeared. In many areas, particularly in the west, a planned countryside of regular fields, straight enclosure roads and new farms had been created. Changes in the landscape on this scale were not experienced again until the mass hedgerow destruction in the 30 years after the second world war. It is clear that while landowners were responsible for some enclosure, especially on the well publicised estates, much of it, especially on the claylands and good loams, was outside their main spheres of influence and was the result of agreements between small owner-occupiers, or of parliamentary acts initiated locally.

Landlords and new farms

The other of Young's improvements which were seen as the duty of the landlord were long leases and large farms. As we have seen, large farms were often created as a result of enclosure, especially in areas of very light soils such as Breckland, and the buildings created stand out in the landscape as a reminder of the landowner's efforts. The group, usually of brick with tiled roofs, consists typically of a square Georgian-style house beside the other buildings, up a long straight road in the middle of its fields. The barn is the largest building, with stables and cattle sheds around a yard in front of the barn, usually to the south. Outward facing cartsheds and granaries complete the group. Straight driftways connect the buildings with all the fields, laid out geometrically across the countryside.

Not all commentators of the time agreed that large farms were necessarily desirable, although William Marshall recognised the change 'as numerous little places of the yeomanry having fallen into the hands of men of fortune, and now being incorporated with their extended estates, are laid out in farms of such sizes as best suit their interest'. The agent at Wolterton saw no advantage in this, stating that 'small farms laid to larger [are] seldom of any advantage to the farm they are let to'. The London agent Nathaniel Kent was not convinced that the productivity of larger farms was necessarily better than that from small ones. He regretted their disappearance, believing that they provided a better living for their farmers than the alternative of labouring on a large farm. 'Agriculture, when it is thrown into a number of hands, becomes the life of industry, the source of plenty, and the fountain of riches to a country; and that monopolised and grasped into a few hands, must dishearten the bulk of mankind.' He also feared that the loss of small farms would mean a scarcity of the products that they traditionally produced, such as eggs, poultry and pig meat. The arguments in favour of large farms were that they encouraged men of capital into farming, they saved on men and horses, and they saved on buildings costs for the landlord.

However, it is not clear how far Norfolk farms did increase in size in the eighteenth and early nineteenth centuries. Again it was on the light lands of the much publicised west, dominated by the large landlords, that the huge farms of over 1,000 acres were found. On Coke's estate there were more farms over 500 acres in 1851 (34) than in 1780 (18) but this was more the result of adding newly enclosed commons to existing farms than of farm amalgamations. Kent thought the ideal size for an arable farm was between 100 and 200 acres, whilst dairy farms could be somewhat smaller. In the fertile areas farms were likely to be smaller than on poorer land. The first census to include details of farm size was 1851, at the end of the period of major land reorganisation. This shows that in the fertile loams over 40% of holdings were under 50 acres, whilst in the west such farms were virtually unknown. However, even in areas where small farms were numerically important, there were also many larger farms so, in terms of acreage, the majority of the land was held by large farmers.

Leases

Finally the provision of long, secure leases was a responsibility of the landlord and was seen by Young as a major characteristic of Norfolk agriculture. We have seen how at Holkham they became increasingly detailed and complex throughout the period. This is a pattern that can be followed on both large and small estates such as that at Thorpe Market and was becoming universal by 1840. The insistence on certain husbandry practices ensured that the land did not become exhausted, but what is not clear is how easy such leases were to enforce. When tenants were in short supply in times of recession, landlords were unlikely to evict tenants for failing to keep to their leases. In times of high prices, there was always a temptation on the part of the farmers to try and take an extra grain crop. As Blaikie wrote, the main aim of the lease was 'to convince a farmer that it is not to his advantage to take *quite* as many corn crops from the land he occupies as it will bear'. From the little evidence available, it is clear that tenants frequently broke the terms of their leases. Crops might fail and a different crop be slipped in. Although cropping books were kept, they were rarely inspected on most estates, and it is likely that in practice the landlord had little control over farming practice on his land. However, most tenants recognised that a system of rotation kept up fertility and yields and so it was to their benefit as well as that of their landlords. 'Improved' agriculture benefited both farmers and landlords, and to promote it they worked together.

Landlords remained key players in agricultural change, particularly in the provision of fixed capital investment, throughout the nineteenth century, and clearly, as rents rose, they were principal beneficiaries. The effects of their efforts on other classes of society will be considered later. Gradually, however, their role in the development of agriculture changed as farming itself developed after 1840.

The farmers

Who were the farmers?

Historians of agricultural history have concentrated heavily on the roles of the landlord, and particularly of the famous names, in the promotion of improved agriculture. We have seen their importance in some areas of the county in creating the appropriate infrastructure or framework for the new agricultural methods to be implemented. Enclosed land, however, did not automatically result in new farming techniques and it was usually the farmer who was responsible for preparing the land by draining and marling and then putting changes into practice. It is to this large and varied body

CHANGING AGRICULTURE

of men and women that we must now turn. There were 6,532 people who described themselves as farmers to the census enumerators in Norfolk in 1851, and there is no reason to think that the number was very different 50 years before. Of these, three and a half thousand farmed less than 50 acres, and of this group, making up half the farmers in the county, we know very little. Only 332 farmed over 500 acres, with the majority of farmers somewhere in between. There must have been a world of difference between the huge capitalist farmers and their small neighbours who can have produced very little for sale. Thus it is difficult to generalise about the farmers as whole. One thing they do seem to have in common: the census figures show that very few farmers had moved far from their place of birth. Most were born in the same parish or a neighbouring one. Even where large farms needed men of capital to farm them, suitable tenants could usually be found locally.

Owners had long been aware of the importance of good tenants to the well-being of their land. As early as 1708 the Guardians running the Holkham estate requested 'that the steward lay a report of the condition of the tenants and their farms in their respective divisions before the guardians for the better distinguishing between the good and bad husbandmen in order to the encouraging of one and the timely admonishing of the other'.

Throughout the period, farmers included gentleman farmers such as Randall Burroughes, who was the largest landowner in Wymondham as well as farming on his own. Then there were large scale tenant farmers such as John Leeds on the Holkham-owned farm at Billingford. These men were well educated. Randall Burroughes' diary contains many Latin tags as well as lengthy quotations from farming and veterinary manuals, while John Leeds labouriously copied part of an English history book into the first 60 pages of his journal as well as extracts from Boswell and Johnson's *Tour of the Hebrides* and Sir John Sinclair's statistical writings. Like Burroughes, he also included extracts from farming and veterinary textbooks. He attended a lecture in Dereham on 'galvanism, electricity etc, explaining various subjects of which I had before but a very imperfect idea'. Lastly there were a huge number of small-scale owner-occupiers and tenants who have left no records of their own, although some were described by Rev. Benjamin Armstrong of Dereham in the 1850s as 'these little masters [who] are often worse off than the labourers'.

A vivid illustration of the varied social make-up of Norfolk farmers is the different types of houses in which they lived. Many of the improving estates

THE FORCES OF CHANGE

rebuilt or extended earlier houses as part of their rebuilding of farms. The attracting of 'men of capital' was all-important and for this a house verging on a 'gentleman's residence' could help. The fine house designed by the well-known architect Samuel Wyatt at Leicester Square Farm, South Creake, in the 1790s looked out over a park-like home pasture to the front. At the back, however, little divided it from the cattle yards and manure heap on which the prosperity of the farm depended. Most houses for farms over 100 acres had two stairways, servants' quarters and up to seven bedrooms. Elsewhere, on the heavier, early-enclosed lands, tenants had to make do with older, albeit often spacious houses, which for centuries had been the centre of prosperous manorial farms and which, particularly in the pastoral areas, provided accommodation for living-in farm servants. Further down

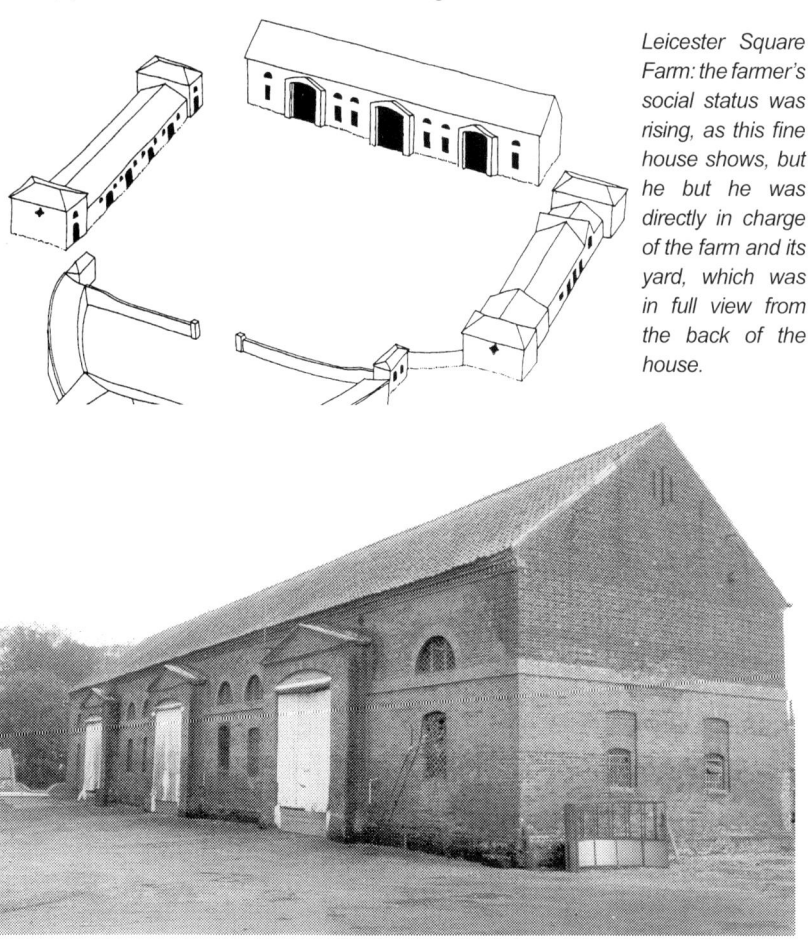

Leicester Square Farm: the farmer's social status was rising, as this fine house shows, but he but he was directly in charge of the farm and its yard, which was in full view from the back of the house.

the social scale, farmhouses were smaller, and on the less well-managed estates, such as that of the Duke of Norfolk in the south of the county, they were often damp and in poor condition.

Probate inventories made on the death of farmers can give some indication of their social status and wealth and confirm the great variety of people who made their living from the land. In 1763, the possessions of the south Norfolk farmer Goddard Frost were valued at the very high figure for the time of £1,263. His extensive farmyard consisted of three barns (one timber-framed, one of stone and one of clay), as well as a riding horse stable, pig yard, stack yard, granary and waggon shed in which there were three waggons and one cart. The importance of dairy cattle is emphasised by the 51 cheeses in the chamber, but no other details of his house or other possessions are given. There were four men servants and one maid servant living in.

Robert Grant of Wimbotsham died in 1772 and left only £274. His modest house consisted of a kitchen with pantry, cellar and dairy as well as a parlour and two bedchambers. However, the parlour was well furnished with seven chairs, a bureau, tea table and corner cupboard containing china and glass. There were pictures on the walls. Peter Dorsey of Terrington St Clements was mainly a sheep farmer when he died in 1792 and there were 85 ewes on his farm. The three cows would have been house cows and the five horses were to work his land. His house had three bedrooms and was well furnished with a desk, clock, three tables, six chairs and two mirrors in his main room.

John Lyall of Dilham was a substantial farmer who died forty years later. His buildings consisted of a wheat and barley barn, stack yard, granary, turnip house, stable with harness for nine working horses, gig house and waggon lodge containing a total of nine waggons and carts. The high social status suggested by his possessions may indicate the increasing prosperity of farmers as a result of the high prices of the Napoleonic wars. Not only did he have a gig but in the 'best parlour' of his six-bedroom house with its servants' attics were 70 books and a barometer. No doubt amongst his books were some of the many texts on agriculture being published in the years following the first appearance of Jethro Tull's *Horse Hoeing Husbandry* in 1731. Many simply reprinted earlier works. Others, according to the historian G. E. Fussell, were 'full of mythical experiments, invented costings and fantastic proposals'. The *General View of the Agriculture of the County of Norfolk* was published in two versions, one by Nathaniel

THE FORCES OF CHANGE

Kent in 1794 and a second by Arthur Young in 1804 and these described the best practice within the county. Sir Humphry Davy's *Elements of Agricultural Chemistry* was first published in 1813 and went into at least six editions. Although rather heavy reading it must have found its way onto many farmers' shelves. Other popular subjects were veterinary and farriery techniques and methods of draining and manuring as well as farmers' calendars giving the tasks throughout the year. A useful ready reckoner, published in Norwich in 1798, was written by John Cullyer of Wicklebrook and went through ten editions before 1851. Nearly all these books claimed to be written by 'practical farmers' of twenty or thirty years' experience.

Arthur Young first commented on the farmers of Norfolk when he visited the county in 1771 and recognised some as the key innovators. True, enclosures by Nicholas Styleman at Snettisham, the interest shown in husbandry by Sir John Turner at Warham and the eccentric methods of Sir Thomas Beevor at Hethel are described, but it is essentially the work of the large-scale farmers rather than their landlords which attracted his attention. This visit was made six years before Thomas William Coke inherited, yet the Holkham tenants like Mr Carr of Massingham, Mr Mallet of Dunton and Mr Billing of Weasenham were already seen as leaders of agricultural improvement. 'You will not among them see the stolen crops that are to be met with among the little occupiers in the eastern part of the county.' To Young, this magic circle of wealthy progressive farmers were of greater significance to farming change than their landlords. As we have seen, Young's view that improved farming was only to be found amongst the large-scale lightland farmers of the north-west did not go uncontested. It could well be that the small farmers of the east understood their good loam soils better than Young, and that in fact it was perfectly possible to take a second cereal crop within a rotation without causing soil exhaustion.

William Marshall in the 1780s attempted a general description of Norfolk farmers which must have been based on those he met in his work as land agent for the Gunton estate. His description therefore is of tenant farmers and especially those with the security of a lease which 'renders them independent. A tenant at will, be his riches what they may, is a subaltern in society.' These were confident, well read and intelligent men. Below them, the 'lower class of Norfolk farmers are the same plain men which farmers in general are in every other county; living in great measure with their servants.'

Twenty years later, Arthur Young was just as complimentary, but again, he only knew those farmers whom he had met, perhaps at the sheep shearings, and

CHANGING AGRICULTURE

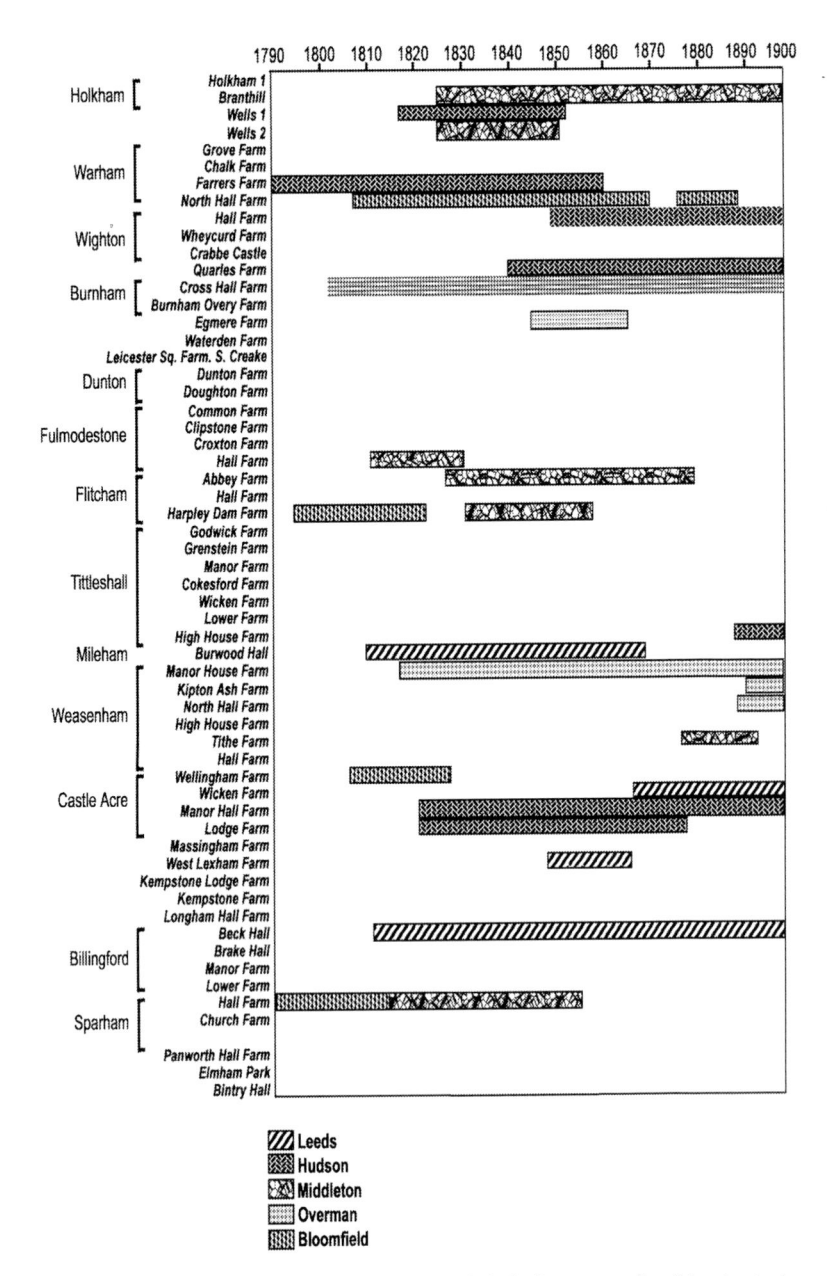

The tenants of the Holkham estate were a tight-knit community; this chart shows how only a few families could control many of the 70 farms.

THE FORCES OF CHANGE

who had entertained him on his visits to Norfolk. These men he described as 'famous for their improvements, the excellency of their management and the hospitable manner in which they live and receive their friends'. He listed 56 men and one woman, both tenants and owners from all parts of the county, who were leaders of their field and had spent time explaining their farming systems to him.

We are fortunate that, from case studies of a few of these men, we are able to put together a picture of the well-to-do Norfolk farmer. Most of them were probably born in the area, and it is clear from the audit books of the Holkham estate that many stayed on the estate for several generations, with different branches of the same family renting farms. They were a tight social community with many marriages taking place within the group. Some owned land as well as renting from the estate. John Leeds, tenant of Beck Hall Farm, Billingford, inherited the 270-acre Oak Farm, Kerdiston, from his father. Another branch of the family owned extensive lands in Reepham and ran a successful tanning business there.

There is no doubt that when estates sought new tenants, one of the most important considerations was whether the candidate was 'a man of capital'. In the 1780s, when farming was depressed, 'landlords and tenants now feel the inconvenience of great farms because of the difficulty of finding those with the capital for such an undertaking'. When seeking tenants for De Grey's Breckland estate in 1822, references were sought for an applicant from outside the area. 'He may bring £1,500 or upwards as his mother and wife's father are people of decent property. He is a working man and a good tenant upon my land; in regard to his principles, I never heard anything that was dishonest.' In 1842 Mr Farrer was thought a suitable tenant because he 'has £800, is not a dissenter and does not care about politics' and another potential taker had 'capital, is a churchman and a conservative'. In 1708, the risk of letting to a farmer without the means to farm successfully was brought home to the guardians of the Holkham estate. A farm at Tittleshall had been 'much abused by this poor fellow, Peake, who . . . was not able to stock or manage it as it ought to have been – if it falls in it will cost £300 to stock it'.

There are a few instances when we gain some insight into the sort of people who were tenants from the surveys undertaken by estate offices. While these usually confine themselves to the farmland and buildings, they occasionally include descriptions of the farmers themselves. When the agent Francis Blaikie visited the farms around Holkham in 1816, he wrote notes on the farmers, especially the younger generation who were the future tenants and whom he sometime took with him on his tours of inspection. He was looking for 'practical' farmers and had little time for book learning. Mr Tuttell Moore of Warham was described as a 'zealous, indefatigable and practical farmer – and no theorist'. Mr Ward, also of Warham, on the other hand 'talks of improving, but at present everything is wrong'. Even at Holkham there were some tenants regarded as 'old-fashioned' and 'backward'. Of a heavy clayland farm in Tittleshall, Blaikie wrote, 'no hope of the present occupier and his son seems of weak intellect'.

Farmers and agricultural change

Cereals

The major achievement of agriculture through the eighteenth and first half of the nineteenth century was the feeding of an expanding and increasingly urbanised population mainly from home-produced food. The staple diet of the English was by this time wheaten bread and so there can be no doubt that as a result of changing agricultural techniques, the output of wheat must have risen dramatically. First, this was through enclosure and the extension of arable. This newly reclaimed land often needed draining, liming and marling and the diaries of the late-eighteenth century farmer Randall Burroughes emphasise the time and physical effort that this work involved. Burroughes farmed in the heavy lands around Wymondham, far away from the accepted heartland of the agricultural revolution in the light soils of north-west Norfolk. Here, what was important was large-scale draining and rearrangement of field boundaries to create larger fields, thus allowing conversion to arable of what had traditionally been livestock and dairying country. However, if the fertility of newly cultivated land was to be kept up, it was necessary to adopt careful farming methods, involving the keeping of animals which provided the only available source of manure. In the early eighteenth century, turnips were grown in small closes to provide extra animal fodder, rather than as part of rotations. As grain prices rose the emphasis was bound to change.

If the landowner's main concern was to keep up the value of his farms both by permanent improvements and by ensuring the maintenance of their fertility, the farmer's was to make as much profit from the produce of the farm as possible. This, as we have seen, could lead to conflict, with farmers in

THE FORCES OF CHANGE

years of high cereal prices, trying to take extra grain crops not permitted in their leases. A 1761 agreement for a heavy-land south-Norfolk farm in Tivetshall stated that the tenant was not to sow any cereal on land which had only been in grass for one year without written permission from the landlord and there are numerous instances in estate correspondence of disagreements between agents and farmers on this issue.

The Norfolk four-course rotation involved alternating cereals (mainly wheat and barley) with break crops (turnips and grasses). We have seen that Young regarded a 'proper rotation', the 'cultivation of turnips, hand-hoed, and 'the cultivation of clover and artificial grasses' as fundamental to Norfolk agriculture. If the turnips were sown in drills rather than broadcast, it was possible to hoe between them to clean the ground. They would all be eaten off in the early winter in the fields by sheep, or lifted and fed in strawed-down yards where manure was made, allowing time for ploughing and manuring to further clean and fertilise the ground before the spring-sown barley crop, thus removing the need for a fallow year. The grass crops included clovers which returned nitrogen to the soil. On poorer soils the grass would be left for several years before being ploughed up and on better soils a second consecutive crop of cereals might be taken.

There is clear evidence both in leases and in tithe books (records kept by incumbents to keep a check on what they were due in tithes) that a great variety of 'improved' rotations containing artificial grasses and turnips were being widely practised by the 1730s. Leases at this early date only indicated in general terms how the arable was to be farmed. Those of the 1720s for the Heydon estate stated typically that the tenant was 'not to sow more than four crops whereof one is turnips before laying down to three years' olland [grass]'. Leases were still not standardised by the time Young wrote his report in 1804, in which 25 pages are devoted to rotations, some of which he approved and some of which he did not. It is clear from these that five- and six-course shifts with two grain crops grown in succession were still common. As the farming of the publicity-conscious light-land estates of north-west Norfolk became regarded as exemplary, then it was the shift most suited to them that became most popular. The extension of arable onto the light soils of former heaths and the extension of drainage on poorer soils made the cultivation of turnips more practical. As grain prices fell after the Napoleonic wars before steadying in the 1830s, farmers were more likely to accept systems where grain made up a smaller proportion of the arable. By 1840 the 'Norfolk four-course' had become symptomatic of good farming practice across Britain. Indeed, hardly had it become

CHANGING AGRICULTURE

accepted before the need for it was superseded by the introduction of the artificial fertilisers and intensive feeding systems associated with the high-input, high-output mid-nineteenth century methods known as 'high' farming.

Livestock

Because wheat was a staple food, the prosperity of farming has long been measured in wheat yields and wheat prices. The virtue of the four-course rotation has been seen in its ability to maintain nutrient balances for the increased output of cereals. An increase in livestock fodder was aimed at increasing the production of manure for the cereal fields. This is graphically shown on the monument to Thomas William Coke, erected in Holkham park at the time of his death in 1842. It is crowned by a wheat sheaf, seen as the ultimate aim of all improvement. Yet, there have been only limited periods in our agricultural history when corn has been king, and this only in favoured areas of Britain which did not even cover all of Norfolk. Throughout successive crashes in cereal prices, in the 1780s, 1820s, 1880s and 1921, animals were the foundation upon which prosperity was rebuilt. William Marshall in the 1780s made it clear that the 'affluent fortunes' that had been made by Norfolk farmers were not from cereal growing, but through 'a superior skill in the purchase of stock and a full supply of money'. This entrepreneurial flair not only involved an eye for a good animal at the right price, but also the ability to sell well at the right time. John Leeds of Billingford rarely missed a Saturday market in Norwich and chose his moments to buy and sell carefully. Not all

The monument to Coke of Holkham is crowned emblematically by a wheatsheaf.

THE FORCES OF CHANGE

rotations were weighted towards cereal production: some produced a large percentage of fodder crops.

Norfolk is traditionally divided by agricultural historians into sheep-keeping and cattle-fattening areas. The heavier soils were used for dairying and bullock fattening and on the lighter soils sheep were kept, firstly on commons and foldcourses, and later, hurdled on turnip fields.

In 1787, Marshall recognised the importance of turnips for feeding cattle, and as this system was only just spreading to other parts of the country he devoted several pages of his report to describing its practice. Cattle were fed in fields or yards or stalls. The advantage of yard and stall feeding was that manure was made, whilst cattle out of doors in the winter tended to puddle or 'poach' the grass. There are plenty of references from the early eighteenth century to putting cattle on turnips. Many leases stipulated that if the outgoing tenant could not agree a price with the incoming one for the turnips standing in the fields after Michaelmas, he could bring his own cattle back in to eat them. Cattle sheds, although a more efficient way of feeding, were expensive to build and few landowners invested in them before the 1790s. Leases on the Heydon estate allowed tenants to cut rough timber for the construction of racks and stalls for cattle. Randall Burroughes described the erection of wooden sheds to the south of his barn to house cattle and the fencing in of yards in 1797, whilst the new farms of the Holkham estate all included brick shelter sheds and yards on the warm south side of barns. Burroughes' cows were kept on grass during the day and in straw yards at night 'till I can buy some turnips for them'. In December 1795 he put 24 bullocks into the yards by the barn where they were to be fed with turnips through the winter.

By the 1780s, some cattle, particularly in Blofield Hundred, were being stalled in sheds of a type more usually associated with intensive methods of 'high farming' than with the eighteenth century. These 'large expensive sheds' consisted of a centre building 36 feet long, 19 feet wide and about 11 feet high with a pair of doors at each end and with a lean-to down each side about 11 feet wide in which the stock was tethered, facing into the centre of the building which was used as a turnip store. The doors at either end could be opened for ventilation. The cattle being fattened were either home-bred as a by-product of dairying, or imported bullocks, mainly from Scotland and some from Ireland. The number of cattle sold at the cattle fairs such as St Faiths increased enormously from the seventeenth century. The first Scottish drovers arrived in September and after sales at Harleston and Woolpit

CHANGING AGRICULTURE

William Marshall described the cattle houses of Blofield in 1787 as 'large, impressive buildings'. The animals were tied in the aisled lean-tos and faced into the centre of the building, where the turnips were stored. Rows of feed troughs divided the turnip store from the animals. Sometimes there was further storage space, probably for hay, above. Very few examples of this building type survive, but they are all found on the broadland edge where rich marsh grazing was available in the summer. The examples shown here are (above) at White House Farm, Upton-with-Fishley; and (below) Manor Farm, Freethorpe.

(Suffolk), they moved to St Faiths for a sale lasting two to three weeks from 17th October. Cattle were also being sold at Norwich, Hempton Green near Fakenham and Hoxne in Suffolk. All business was concluded by Christmas. Most cattle bought were four years old and would be fattened for one winter before being sold. Younger stock were bought by those with summer pastures who could afford to keep animals longer. Although turnips were the most important winter feed, cabbages were also grown, particularly on heavier soils. Six thousand were planted on the home farm at Earsham beside the River Waveney in 1793 and gangs were paid for hoeing both turnips and cabbages. Gradually mangold wurzels replaced turnips as they were more resistant to disease, and artificial feeds such as oil cake were being bought in as early as the 1770s on some of the large farms such as Mr Carr's at Massingham.

Once the stock were ready for sale, the farmer had several options open to him. As they came fat, Randall Burroughes sold to the local butcher from Norwich or Wymondham. Selling locally could be a long process. On 28th February 1796 Mr Lowden, a Norwich butcher, along with a horse dealer, came to his farm to buy stock, negotiating with Burroughes' farm manager. 'They continued together from 11 o'clock morning 'till near six in the evening, drank three bottles of port wine & two of ale but could not agree upon terms'.

The development of a more commercial approach to agriculture was one of the stimuli of innovation, but it was only possible where sophisticated marketing systems were in place. The selling of beasts further afield was also a complicated business which involved a week-long walk to London. Marshall reckoned that between three quarters and two thirds of Norfolk bullocks were sold at Smithfield by 1787. Drovers set out from St Faiths once or twice a week and were met by the salesmen at Mile End. Burroughes took four bullocks to Tasburgh to meet the drovers on Sunday 17th April 1796 and on 19th he received an account from the salesman. He made £18 15*s* each – he had hoped for £20 'but the market was reported to be very much overstock'd & had they not been well fatted would have return'd a very bad account.'

Sheep had formed an essential part of the Norfolk farming system from the Middle Ages when wool had provided one of the bases of the county's wealth. This importance continued throughout the eighteenth and nineteenth centuries, and the maxim 'small in size but great in value' is recorded under the sculpture of a Southdown sheep on the monument to Thomas

CHANGING AGRICULTURE

William Coke in Holkham Park. Breeding was limited mainly to the light soils of the north-west, but farm sales evidence shows that they were kept far more widely and that Marshall's statement that north-east Norfolk was 'as free from sheep as elephants' was an exaggeration. In fact they were kept on the good loams as well as the heaths of the east, and provided the raw materials for the industries of the Worstead and Aylsham area. Cawston, along with Kipton Ash, near Fakenham, were host to the largest sheep sales in the county. That at Kipton was held on the common edge at the end of August when the breeders from the west sold stock to the graziers. 'This is entirely a fair of business: scarcely a woman or a townsman to be seen in it. Many of the first farmers in Norfolk were there today [28th August 1876]'. The tithe accounts for the heavy-land parish of Mattishall show 200 ewes and 800 'sheep' being kept there in the 1790s and there were normally about 100 sheep on Randall Burroughes' Wymondham farm.

As we move onto the light lands sheep become much more important. The foldcourse system is uniquely East Anglian and enabled light lands to be manured. The eighteenth century saw a shift in methods of sheep management and thus a change in the type of sheep kept. First, extensive systems of feeding across heaths and foldcourses were replaced by an intensive system of hurdling tightly on turnips and grazing on improved pastures. The largest flocks of up to 1,000 sheep remained in the west of the county, particularly in Breckland. Traditionally it was the black-faced Norfolk Horn sheep that dominated, but this leggy animal, which was said to survive on the poorest of grazing, was slow growing, and the original reason for the inauguration of the Holkham sheep shearings was the promotion of more fast-maturing breeds, first Bakewell's New Leicester and later, from

The native sheep of Norfolk was the black-faced Norfolk Horn, shown here in an illustration from Nathaniel Kent's book on Norfolk agriculture (1796 edition). Through the early nineteenth century it was gradually replaced by the faster maturing Southdown. However, half-breeds were also popular, and by the 1860s the Norfolk cross Southdown was recognised as a new commercial breed – the Suffolk.

THE FORCES OF CHANGE

1792, the Southdown. After 1806, Coke decided that Southdowns were preferable to Leicesters and he sold his flock. Interest in them increased as more feed, in the form of turnips, became available and some huge flocks were kept in the west of the county. A flock of nearly 2,000, including 581 lambs, was sold at West Tofts in 1823. The breeding of pure-bred pedigree sheep was very much a gentlemanly pursuit with prize rams exchanging hands at extravagant prices, and the ordinary farmer had to content himself with cross-breds. Rising from nothing in the 1790s, half-breds came to dominate the market at dispersal sales across the sheep keeping areas of the county. However, these in themselves were to become an accepted 'improved' breed with the Norfolk/Southdown cross becoming registered as the Suffolk breed in the 1860s.

Sheep were more important than cattle for producing manure in Breckland and the north-west, partly because the light soils did not produce enough straw for cattle yards and partly because, even after enclosure, there were still large areas of open sheep walk. Land owners appreciated the value of sheep, and leases often stipulated the number to be kept on the newly enclosed land. Some 500 ewes were to be kept on the 920 acre Manor Farm at Heacham, and at Tottington in 1774 flocks of not less than 500 sheep should be kept. Newly planted hedges were to be protected from them by hurdles. Here it was recognised that the 'improvement of the light lands depends on a due quantity of sheep being kept to fold upon them'. The problem was that leases also laid down the quantities of marl that should be spread to fit these light soils for cereals. Up to 75 loads per acre were to be applied during the first four years of the lease, dropping to 50 and then 25 loads for the rest of a 12 or 18 year lease. Tenants claimed that this was changing the mineral balance of the soil and causing the ewes to abort (or warp). In 1836 Mr Lincoln claimed that he had lost 200 lambs 'being warped in consequence of claying land'. His lease was renewed in 1845 and Lincoln chose to ignore the marling clauses. Tenants knew that sheep were the backbone of Breckland farming and were not prepared to take the risk of harming their flocks, while the landlord was not willing to push his luck by forcing tenants to fulfil obligations set out in their leases.

It can be seen that livestock husbandry did not play a secondary role to arable farming. Rather it was an essential and truly integrated part of the farming system. In the east cattle provided an important cash crop for farmers, as did sheep in the west. A highly developed marketing system existed which is in itself an indication of the importance of stock to the farming economy. Changes in livestock husbandry in the eighteenth century were

CHANGING AGRICULTURE

as great as those in the arable sector. As far as sheep were concerned folding on stubble and grazing on heaths was replaced by feeding on turnip fields and temporary leys. The type of sheep kept necessarily changed to suit the new situation. Cattle management too was changing as the animals were increasingly turnip-fed in yards rather than in the fields. Gradually too new foodstuffs were coming in to break the closed-circuit system of the four-course whereby food and fertilisers were all produced on the farm, with only stock for fattening and seed being bought, and fatstock and cereals being sold. This is a change which we shall see being further developed during the high farming period after 1840.

The period 1720–1840 was one in which three generations of farmers across Norfolk were responsible for fundamental changes in the farming system as well as, by draining, marling and hedging, transforming the soil and the landscape itself. We have already seen that as far as the landlords were concerned this was a very profitable period, with rents rising often by 100%. How profitable it was for the farmer is far more difficult to judge. Few kept careful accounts and when they did produce figures, it was usually to prove a point. From the correspondence of agents, it is clear that there were years of hardship during the eighteenth century when rents were difficult to collect. In the 1780s the farmers on the light lands suffered from two dry summers and a severe winter as well as 'insufficient crops, the low price of several grains and the ambition of hiring great farms which they are unable to stock without borrowing money', so that in 1786 they were worth 20% less than three years earlier. The tenant at Langford had failed and there were 27 sales of farmers' stock advertised in the Norwich paper.

By the 1790s the situation had improved and the years of the Napoleonic wars saw wheat prices rise to previously unheard-of levels. Landlords pushed up rents, but still farmers did very well. As soon as the war ended confidence collapsed and with it came problems for some farmers. In 1815 Mr Sewell of Thetford did not pay his rent until the last minute: 'On Monday last it was my intention to have written to you to have distrained Mr Sewell's wheat stacks for a whole year of rent due – but he has promised to pay'. In 1816 De Grey's tenants requested rebates and the agent recommended reductions on small farms which had been over-valued. Away from the poor Breckland soils, farmers, particularly the larger ones, were more able to weather depression. John Blythe of Burnham told the Board of Agriculture Enquiry into the State of Agriculture in 1816 that because farms in his area were large, their farmers were greater capitalists and so when rent was paid it was from other resources such as cutting down on stock, which in

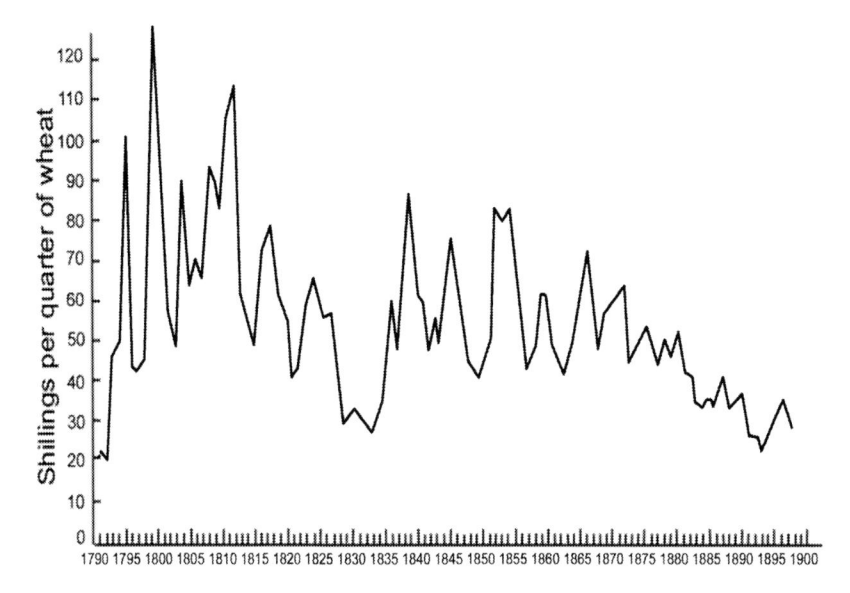

Graph showing the price of wheat (in shillings per quarter) in Norwich market in January of each year. The high prices of the Napoleonic wars contrast with the slump of the 1820s, steadying in the 1840s and 50s and declining after 1870.

the long run was bad for the farm. Not everyone agreed that large farmers were necessarily better off. In 1804 Rowland Hunt wrote in *Communications to the Board of Agriculture* that 'a farm of fifty acres makes many a pauper; a farm of a thousand acres makes many a bankrupt' and he thought that 150–300 acres was the best size for a farm.

The post-Napoleonic war depression forced some farmers to give up with the reason for sale given in the newspaper announcements often including phrases such as 'changing business', 'leaving his farm for a mercantile occupation' or simply for 'the benefit of creditors' and in 'distress for rent'. In 1822 the well-respected Holkham tenant and breeder of Southdowns, Mr Purdy of Castle Acre, was forced to sell and in 1823–4 the number of farmers giving up or moving to smaller farms rose. The amount of rent collected on the Holkham estate dropped and rebates of up to 30 per cent were agreed. A second period of depression occurred in 1826 following the banking crisis of 1825–6 and the failure of Messrs Day's bank in Norwich, which in its turn brought ruin to some of its farming customers. The general gloom also affected the price of grain; John Leeds recorded in his journal in December 1825, 'The price of grain now very dull, as well as the general state of things, from the extreme depression in the money market, the funds have fallen by 10 per cent and the mercantile and manufacturing interests equally unhinged and without any apparent cause at present assigned.' Prices remained unsettled throughout much of 1826, with a recovery beginning in 1827. However, many farmers were still finding times difficult into the 1830s; Mr Margitson, agent for the Earsham estate, described the tenants as 'an industrious and frugal set of men . . . and if they do not get on, it is from the times and seasons working against them'. Arrears were still high on De Grey's estates in 1835 because of 'late bad years'.

It is clear that in spite of improved yields, farmers were still very vulnerable to moves in prices, often brought about by financial speculations outside their control. As is to be expected, it was the large-scale farmers who were better able to weather depressions. Landlords, because more of their capital was now tied up in the land as a result of improvements carried out, now expected a higher percentage of the farming income in the form of increased rents. This in turn put further pressure on the farmers, reducing their ability to find the money needed to stock their intensive enterprises. By 1840, however, farming had generally recovered from the years of depression and was entering a new and different phase of development known to contemporaries as 'high farming'.

The labourers

The new agricultural systems of the eighteenth century relied heavily on a large cheap labour force. With more land brought into cultivation, there was a shift to arable from pastoral farming. Turnips, a vital part of all rotations, were a labour-intensive crop, so it was likely that more rather than less labour would be required. It has been calculated that there was a 45% increase in the amount of labour needed to operate a four-course rotation above that needed to work a three-year cycle in which one third was fallow at any one time. The old three-field system was not typical of Norfolk at the beginning of our period, but calculations such as this do emphasise the importance of an abundant labour supply to the new farming methods. Unlike the 'industrial revolution' the 'agricultural revolution' was not led by machines but by hand labour, and Norfolk was in a good position to provide that labour. Here was a county whose economy had been industrial and commercial. Industry had been based on textiles and the many prosperous market towns had grown wealthy on this trade. However, the factories, first water- and then steam-powered, that were springing up along the fast flowing rivers and near the coal fields of the north and west of England and southern Scotland were stifling the textile hand-workers elsewhere. A government report of 1834 revealed that only in a few places was the textile industry still employing workers. In Shotesham children filled bobbins for weavers and in Sprowston and Worstead there was still some weaving, 'but no constant employment'. Elsewhere, such as at Brockdish, 'spinning is hardly worth doing, a few are employed in making lace, straw hats sewing and weaving'. In Saxlingham spinning was no longer important.

In the countryside a surplus of labour led not only to unemployment but also to starvation-level wages. William Marshall noted in 1787 that 'a Norfolk labourer will do as much for one shilling as two men in many other places will do for eighteen pence each'. The situation had not changed by 1804 when Arthur Young reported that 'the circumstance in rural economy which for many years distinguished Norfolk in a remarkable manner was the cheapness wherewith the farmer carried out his business'. Labour was not only cheap, but had 'greater activity and a spirit of exertion'. Nathaniel Kent recorded that prices were 60% higher than they had been in 1750 whilst wages had only risen by 25%. Inevitably, this situation erupted into discontent and sporadic rioting, particularly in the years of grain shortages and high prices. Burroughes' journal in 1795 records a 'deputation of labourers' calling at his brother's farm in Burlingham demanding a rise in wages. Anticipating trouble, he had taken the precaution of calling up the local yeomanry and alerting the military, but in fact the meeting was a peaceful one. There is no indication in the diary as to whether higher wages were agreed. Later in the same year there were disturbances in Hethersett where labourers found grain in store that they thought was being hoarded against further rises in prices. They demanded that it should be taken to Wymondham and distributed amongst the poor. Peace was finally restored by the local magistrates giving out money for food.

After the war the problem of unemployment was exacerbated by a slump in industry and by returning soldiers. Discontent was endemic, the most well-known disturbances being the bread riots around Ely and Littleport in 1816 which resulted in the hanging of five men outside Norwich Castle. Although grain prices fell in the following years and actual starvation was less of a threat, unemployment remained a problem. As late as 1831, census figures suggest that an eighth of the men in rural parishes were out of work. However, farmers argued that this was a surplus they needed at some times of year. At harvest they wanted a pool of labour they could call upon.

CHANGING AGRICULTURE

With poverty and underemployment at this high level and the population in the countryside continuing to rise, there was little incentive for farmers to mechanise. Indeed, where threshing machines were occasionally installed, they were sometimes worked by men on a treadmill rather than being turned by horses, which, according to a Costessey farmer in 1843, 'costs as much as to flail – we are too many men so I do not use horse power'. It was not until the population of the countryside began to level out in the 1850s as more people moved to the towns that wages generally began to creep up.

It is against this background that the role of the labourer in the development of Norfolk agriculture must be seen. The availability of cheap labour for such labour-intensive activities as the yard feeding of stock, turnip hoeing and lifting, digging, carting and spreading of marl and the clearing, carting and spreading of muck was essential to the successful operation of the new farming systems.

It is very difficult to assess the number of labourers employed on a farm, partly because different types of labourers were costed for in different ways. Farm labour books often only cover regular day labourers or the yearly hired servants, and not those hired individually or in gangs on piece rates. Divisions could also be blurred. For instance, then as now, farmers of small acreages could sometimes work for their larger neighbours. Randall Burroughes' ex-coachman rented land from him, but occasionally worked for him as well. The size of the farmers' family would influence the number of workers employed, particularly on smaller farms. One of the questions asked in the 1834 report on the operation of the Poor Law was 'What was the number of labourers sufficient for the proper cultivation of the land?' The information was provided by church wardens and most left this question out. At Brockdish one labourer for every 30 acres was suggested, while at Starston the figure was one labourer and one boy for every 100 acres. At Downham Market the church warden was less sure and reported that 'the number varied according to the good or bad mode of cultivation'. Similarly, when the journalist Richard Noverre Bacon was researching for his book on Norfolk agriculture (1844), he asked farmers how many labourers they employed, but the question was only vaguely answered, or left out altogether. Henry Blyth of Burnham Deepdale pointed out that one of the differences between farming now and farming thirty or forty years ago was that more manual labour was employed. A farmer of 400 acres employed 'upwards of 20 men' while another stated, 'just as many men I want for exactly the time it takes to do the work'. Here we see the nature of much

FORCES OF CHANGE

farm work; it was insecure and part time. All the correspondents said they 'put out' as much work as they could which emphasises the difficulty in coming to any firm conclusions about the number employed, and thus the labour productivity at the time.

However, figures in the 1851 census represent the situation just before the introduction of machinery would have influenced the number employed. They suggest that the average number of acres per labourer on a medium-sized 80 to 200 acre farm was between 20 and 30. In 1853, the Suffolk agricultural writers Hugh and William Raynbird in their revised edition of Rham's *Dictionary of the Farm*, published in 1853, suggested that 16 labourers and 8 boys would be employed on a 'typical' 400-acre mixed heavy-land farm.

Types of labour

Those with the most security were the yearly hired men who had traditionally been given their keep in the farm house as well as a wage. Their work usually involved animals, working with horses (horsemen, ploughmen and teamsmen), with cattle (yardsmen) or with sheep (shepherds). By the 1790s living in was less usual and instead they were given cottages or an allowance. When John Smith left Randall Burroughes' household in 1795 there was only one 'hired servant left in the family'. A witness to the Board of Agriculture enquiry of 1801 pointed out that living in was declining just as food prices were rising. 'You cease to feed your men when it is hardest for them to feed themselves.' Young reported that living in was declining in many areas by 1804 and instead an allowance was paid. The result was, according to Young, that when the farmer no longer had his workers under his roof, the Sabbath was neglected. Average rates of pay given by Nathaniel Kent in his survey of Norfolk agriculture of 1796 range from nine to ten guineas for a head carter to eight pounds for a yard man, plus lodging at between seven and eight shillings a week. This figure is similar to that paid by Burroughes, but higher than that in Ringstead where some yearly labourers earned as little as five pounds a year.

Below the yearly paid men were those on day or piece rates, and it is very difficult to establish their rates of pay, but figures in the region of 1s 6d a day or 10s a week seem to have been usual by the 1790s, although individual figures from different farms can vary considerably from this norm. The 1834

CHANGING AGRICULTURE

survey suggested £30 a year as an average income, whilst in East Rudham 'able labourers who thresh the great part of the year earn between 14/- and 17/- per week'. Arthur Young and the contributors to his publication, *The Annals of Agriculture*, calculated figures for the annual income and expenditure of labourers which showed that even with wages coming in from more than one member of the family, it was almost impossible to remain out of debt. A spell of illness or unemployment left a family with nothing but poor relief to fall back on.

The problem with these figures is that they take no account of, on the one hand, the help in kind which a labourer might expect to receive, or, on the other, how long might be the periods of unemployment suffered. Help in kind had always played an important part in cementing the master–man relationship in farming. Randall Burroughes sold his labourers meal at a lower than market price and in the particularly bad years of 1794–5 the farmers of Fincham were operating a similar scheme. Meals were also provided by Burroughes when labourers had worked together for a day to bring a stack into the barn for threshing, as well as during harvest. However, the harvest meal was not always acceptable. In September 1799 the men 'refus'd to eat the pluck of the pig alleging it did not agree with them. This I consider'd an idle excuse & resolv'd to consider it unreasonable daintiness and to treat them accordingly'. The farmer was expected to provide the food and drink for a 'frolic' when all the grain was in. Gleaning was also an important source of food for labourers with four to six bushels being collected by some families. The advantage to the farmer of giving perks rather than raising wages was that it emphasised the paternalistic nature of the master–man relationship. It was always much more difficult to reduce wages if prices fell than to withdraw gifts and allowances in kind.

It was only those regularly employed who could hope for this sort of paternalism and, as there was no shortage of labour, the farmers were able to organise much of their work on piece rates. On Randall Burroughes' farm, the piece rates were generally set so as to allow a minimum wage of 1s 6d a day to be earned (if the labourers worked hard), and this seems to have been standard on many farms. Hoeing, threshing, sowing seed, ditching, marling, planting hedges and digging drains was typically paid for by piece rates. It has been calculated from the 1831 and 1841 censuses that there were three casual to every two regular workers in Norfolk at the time. Some of these workers, particularly women and children, worked in the notorious gangs operating from the so-called 'open' villages such as Castle Acre in north-west Norfolk where gang masters were able to exploit the labour surplus in these large settlements against the shortages on the neighbouring landlord-controlled farms where no cottages were provided for labourers to live near their work.

The working year

However the work was organised, the same seasonal rhythm of activities took place on Norfolk farms, both large and small. Autumn and winter work included muck spreading, ditching and hedging followed by ploughing, firstly for winter-sown wheat and then for spring crops. Carting took up an enormous amount of time and energy before the railways. Clay and muck was taken out into the fields, and stones and root crops taken off. Corn was

Norfolk plough, as depicted in Nathaniel Kent's book on Norfolk agriculture (1796).

taken to market and coal brought back. As the turnip lands were cleared they too could be ploughed, as could any fallow and stubbles which had been grazed. There were usually several ploughings to get rid of weeds before a final harrowing to produce a seed bed. Throughout the winter, the corn would be threshed to provide both a steady supply for sale and straw for the cattle in the yards and chaff (chopped straw) to be mixed with oats as fodder for the horses. With the spring came the main sowing season, firstly the barley and oats followed by the turnips in early summer. Spring cereals were often undersown with grass which would grow up amongst the stubbles to provide pasture and a hay crop. By the early summer the livestock would be outside grazing on the meadows and then, after the hay harvest, on the aftermath there. By May the winter wheat needed hoeing and weeding and the newly-sown crops, rolling. Docks needed cutting and thistles pulling – work that was often undertaken by gangs of women and children. The busiest time of year was the summer with first the hay and then the grain harvest. Alongside this the turnips needed hoeing, usually twice during the summer.

CHANGING AGRICULTURE

All workers, including day- and yearly-hired men, were involved together in the grain harvest for which a separate 'harvest bargain' was made. In 1795 Burroughes employed 10 men and a boy at 7s 6d an acre which amounted to about two guineas a man. William Marshall calculated the price of a harvest man as between 35s and 40s plus board. Harvest usually lasted between four and six weeks and the harvest bargain often included the hoeing of turnips when the crop was too wet to cut. The Castle Acre farmer John Hudson worked an enormous farm on which there were 600 acres of cereals to harvest. He described the management of his harvest to Richard Noverre Bacon in 1844, shortly before mechanisation arrived on the harvest field. A hundred to 120 men, women and children were hired. There were 34 mowers. Each man was followed by two women, or one woman and a child, to gather up the corn into small sheaves, which they placed ten to a shock. The stubble was then horse-raked, and the rakings tied up. Three hundred acres of wheat could be cut in six days, but the carting back to the farm yard took another eight. When the weather was favourable, harvest could be completed in three weeks.

Above: Using a horse to draw a rake in Norfolk. The horse rake pictured left is one of the agricultural implements offered for sale by Garrett's of Leiston (Suffolk) in the 1850s.

The most time-consuming tasks on the farm were ploughing and the hoeing and pulling of turnips. Indeed it was only the availability of so much cheap labour that made the culture of turnips viable, allowing the rotations associated with the initial intensification of agriculture to be implemented. Bacon wrote in 1844, 'Upon no other crop is so large an outlay made either in manure or labour.' Not far behind was the time taken in carting, spreading muck and sowing seed.

The changing methods of work

Although machinery only found its way very slowly onto Norfolk farms, there were improvements in the hand tools used that increased the productivity of labour. First there were changes on the harvest field where scythes were replacing the short-handled sickles. This allowed the crop to be cut, or mowed, much closer to the ground so that long straw was available for bedding cattle in yards, and small quantities for thatching. Reaping with sickles was not only more back-breaking, it was also slower. Between a third and half an acre could be reaped in a day whilst an acre could be mown. The stubble was left much higher when the crop was reaped. This was fine if animals were to be put onto the stubble to eat what they could forage and trample the straw into the ground and thus manure the field, but was not so suited to the current systems where animals were inwintered. Generally the change from reaping with sickles or reaping hooks to mowing with scythes had begun by the 1790s, but it was slow, with some farmers still reaping part of their crop in the 1840s. The 1834 report lists many parishes where reaping was still usual. Arguments in favour of reaping were based on the belief that the higher stubble, when worked into the ground by ploughing, was good for the succeeding crop, whilst mowing was 'injurious to the next crop'. Other farmers claimed to have tried both methods and could see no difference. A simple prejudice in favour of old-fashioned methods is betrayed in remarks such as those of the Billingford farmer John Leeds who in the 1820s recorded that he had mowed some wheat, 'a slovenly way in my opinion'.

There was little incentive to adopt labour-saving methods when there was a surplus of labour and putting men out of work merely increased the poor rates. This attitude is clearly shown in farmers' attitudes to the seed drill. Jethro Tull's seed drill, publicised through his book *Horse-hoeing husbandry* in 1733, is often seen as a key development in the agricultural

CHANGING AGRICULTURE

revolution because it allowed for the sowing of seeds in lines so that hoeing could take place between the rows. Many improved models were available to farmers by the end of the eighteenth century, but William Marshall reported that drilling was 'entirely unpracticed' in Norfolk in the 1780s. This was not because farmers were still broadcasting their seed and ignoring the need to hoe, but rather because they had taken to the extremely labour-intensive methods of hand dibbling and this remained the favoured method on many farms up to the 1840s. Drilling was said to waste seed and it was more difficult to get into the corners of fields with a machine than with a man using a pair of dibbles. A man with three droppers (usually women or children – 'although the dropping of seeds is often very improperly done by the children') could dibble about half an acre and at a cost of about 7*s* 6*d* an acre in the 1840s, and this was thought worth while. Gradually however, from the 1790s, seed drills were finding their way onto Norfolk farms.

Arthur Young gave as a reason for the writing of a second report on Norfolk agriculture in 1804 closely following that of Kent in 1796, the rapidity with which the practice of drilling was spreading through the county; but he only noted a handful of farmers who were drilling their corn by 1804. Cooke's seed drills begin to appear in Michaelmas sales advertisements from 1800. One enthusiast claimed that a drill could do as much in an hour as a man with two dibbles in a day. However, change was slow, some of Bacon's correspondents reporting in the early 1840s that they dibbled wheat and drilled barley. It was not until the labour surplus finally declined in the middle of the century that this time-consuming method of sowing was finally largely abandoned.

Seed drill.

The most important mechanical innovation of the late eighteenth century was the threshing machine. It was invented in Scotland and was available from the 1780s, but very few found their way into Norfolk. A public notice in the *Norfolk Chronicle* in 1804 announced that one was to be erected by J. Ball on the farms of Mr Gee near Norwich and Thomas Sepping at Whitehall near Syderstone 'where there will be demonstrations of them at work'. None found their way into dispersal sales until 1811, when several, varying from one to four horsepower and all worked by horse-power, are listed. One

From the 1780s onwards machinery (to begin with, mainly threshing machines) could be powered by non-human motive power. Horse engines were the most common in Norfolk, although water power was also used. These engines were usually housed in round houses and only a few survive within the county. This horse engine survives within a round house in South Creake.

in Thompson was advertised in 1813 as capable of 'clearing from the straw 20 to 40 coombs (80 to 120 bushels) of corn in 12 hours by the power of two to four horses'. This compared with one at Rudham inspected by Arthur Young in 1804, where four horses powered an engine which threshed 40 coombs (160 bushels) of wheat, 50 coombs (200 bushels) of barley or 60 coombs (240 bushels) of oats or peas in eight hours. Others seen by Young were made by Mr Wigful of Lynn who had taken out a patent in 1795 for a large machine which would have cost £100 to buy, whilst a few came from Scotland. All those described by Young in 1804 were to be worked by horse-power, although a steam engine to be used only for 'agricultural purposes' was being erected at Heydon. The gearing for the horse engines was housed and at Rudham there was a granary over the horse wheel. These engine houses were normally round or hexagonal and one possibly eighteenth-century roundhouse survives at Plumstead near Norwich. With a diameter of over 40 feet, it is far larger than the later example at South Creake, but according to Young some of these early examples needed as many as eight horses to power them, so a large building would have been needed. Although Young produced figures to show that grain could be threshed more cheaply and cleanly by machine, the initial but very patchy

CHANGING AGRICULTURE

The threshing machine was a significant technical breakthrough of the 1780s, but it was unpopular with labourers who would suffer winter unemployment if hand-flailing came to an end. Few machines found their way into Norfolk before the 1830s.

interest in them died out after the boom years of the Napoleonic wars.

This lack of enthusiasm is in great contrast to the speed of uptake in the Lothians of Scotland, the other region of the United Kingdom famous for agricultural progress, where, unlike Norfolk, there was a shortage of labour. Indeed, so worried were Norfolk landlords about the surplus of labour that they made efforts to prevent the spread of machines. A lease for a farm in Snettisham, granted in 1817, stipulated that the tenant was to stack the corn in the barn and stack yards and 'there to thrash the same by men with hand flails'. Fears that machines would result in a lack of winter employment resulted in an outbreak of machine breaking in 1822 and landlords continued to prohibit their use through leases. A printed form for standard leases granted by the Langley estate in 1829 included the stipulation that the tenant was 'not to affix during the said term any threshing or other machinery to any building'. A more serious outbreak of machine breaking occurred in 1830 and was known as the Swing riots. From the middle of November, for nearly two months arson and machine breaking affected over 150 Norfolk parishes. These attacks were often preceded by a letter to the farmer signed by 'Swing'. In an attempt to stem the troubles, North Walsham magistrates printed a notice urging farmers to dismantle threshing machines and increase wages. Although sporadic machine breaking continued into the 1840s, the worst of the problems of a labour surplus seems to have been over, at least in some parts of the county, by the mid 30s. The agent of the Earsham estate wrote to the owner in 1835 stating that 'Chambers has made several applications to me for permission to use a threshing machine. He says they are very generally used in the neighbourhood and as there are no men unemployed he hopes to be allowed to use one occasionally'.

The unrest of the 1830s was seen by farmers as having a serious effect on the relationship between the farmer and his men. The old paternal system which is so evident in Randall Burroughes' diary was replaced by suspicion and fear. When witnesses were asked by the 1833 Parliamentary Enquiry into the State of Agriculture whether there was the same good feeling between farmers and labourers as formerly the replies were clear. One witness said, 'Nothing like before the fires', whilst another reported that 'the struggle of keeping profits up by beating down wages is so painful that men are indisposed to embark on it'. A third commented on 'the nuisance of having a quantity of unemployed people teasing for allowances and threatening'.

There were other reasons for the slow uptake of threshing machines. Some farmers thought that malting barley was bruised and the straw broken. If the yield was good, a man could thresh a quarter (8 bushels) a day and when the cost of hiring a machine was taken into account it could be just as cheap. But if it was not much cheaper, it was certainly much quicker to use a machine which could allow a farmer to take advantage of sudden swings in the grain market. However flails continued to be used on small farms and also by labourers threshing out their gleanings well after 1850.

It is clear that Norfolk deserves its reputation as a county where new farming techniques were replacing traditional cropping patterns and increasing yields. Having a cheap workforce allowed the implementation of rotations including the labour-intensive turnip; this was perhaps of greater importance than the more frequently recognised role of the improving landlord. The lack of alternative employment in a de-industrialising economy allowed the county to become the initiator of agricultural progress. The agriculture of other regions such as Northumberland and the Lothians of Scotland, which became noted as seats of improvement, differed from that of Norfolk in that they had industries and mines close by, so they were short of cheap labour; machinery, particularly threshing machines, was developed early in such areas and spread fast. Norfolk fell far behind in this respect and its workforce remained one of the lowest paid in the country into the twentieth century.

HIGH FARMING c.1840–1875

The middle years of the nineteenth century saw a nation-wide interest in agricultural development in which Norfolk played a significant, if not pioneering, role. As depression set in during the last quarter of the century, contemporaries looked back nostalgically to the golden mid-century years of 'high farming' which had provided a good living for farmers. Evidence for this prosperity can be seen in the countryside today. Not only were new and substantial farm buildings erected to house the increasingly valued livestock, but the older Georgian houses were extended with new porches and bay windows, displaying the social status of the farmer.

The term 'high farming' is difficult to define, but it was in current use by the late 1830s, when one of the questions addressed by the compilers of the tithe files was to note whether a parish was 'high or low farmed so as to affect materially the quantity of produce'. The good loams of Sloley were said in 1837 to be high farmed, whilst on the light lands of Stanford in Breckland it was only high farming that allowed corn to be produced. In Mattishall, on the other hand, the lack of leases was said to prevent high farming. 'The tenants are in moderate circumstances and not of that capital to use an expensive or speculative husbandry.' These three replies from Norfolk parishes give a clear indication of what contemporaries meant by high farming and where they expected it to be found. It was a high input, high output system relying largely on the new chemical fertilisers and imported animal feeds and suited to large farms (over 300 acres) where the tenants had secure leases and access to capital. This was particularly on the poorer land (the 'hungry' soils) which could only be profitable if well manured. It was therefore on the light soils of the west that high farming was most likely to be found. The Norfolk farmer and commentator Clare Sewell Read wrote of west Norfolk in 1858 that there were no other areas where 'such an amount of the necessities of life are raised by artificial manure'.

Enthusiasm for high farming really took off in the 1840s. The Royal Agricultural Society was founded in 1838 with the motto 'Practice with Science' and the first volume of its *Journal* appeared in 1840. Writing in 1843, the secretary of the Society, Philip Pusey, defined 'high farming' as a system in which the use of oil cake for feeding livestock, chemical fertilisers on the crops and drainage tiles under the ground were the main elements. The Royal Agricultural College at Cirencester opened in 1842 and gradually attitudes to farming education changed. No longer was it sufficient to be a 'practical' farmer. An understanding of science, particularly soil chemistry, was seen as important.

Whilst there is no doubt that interest in this capital-intensive farming reached a peak in Norfolk and elsewhere in the 1850s and 60s, its origins are more difficult to trace. In Norfolk, many of the elements associated with high farming can be found by the late eighteenth century. High farming demanded heavy investment on the part of the farmer, but there was nothing new in this. Marling and stocking had always been expensive and landlords had always sought 'men of capital'. In times of agricultural uncertainty it was difficult to attract these tenants. A witness to the 1833 Parliamentary Enquiry states that men with £4–5,000 to invest were more ready to put their capital into trade than farming. Gradually confidence returned, particularly after the repeal of the Corn Laws in 1846, which allowed free trade in grain and did not have the disastrous consequences predicted. By mid-century, the amount of capital needed by the high farmer was certainly more than that required by his father and grandfather. Richard Noverre Bacon's Norfolk report, published in 1844, contained figures for a light-land farm which showed that the cost of cultivation had risen from £1 15*s* an acre in 1790, to £3 11*s* in 1820. Figures of £4 an acre seem typical of the 1840s, perhaps reaching £6 to £9 by 1860. The costs on the light lands were greatest. The land agent letting a farm in Sturston in 1842 wrote, 'To farm with spirit and improve it as now offered will require nearly £5,000.'

There were some ways in which the landlord could enable the tenant to make the most of the more intensive farming methods, whilst at the same time sharing in the profits which they brought. He could improve the drain-

CHANGING AGRICULTURE

age of the land, thus bringing more farmland into arable production, and he could provide the buildings needed for more intensive livestock production. For both these investments he could charge a percentage increase on the rents.

Farm buildings

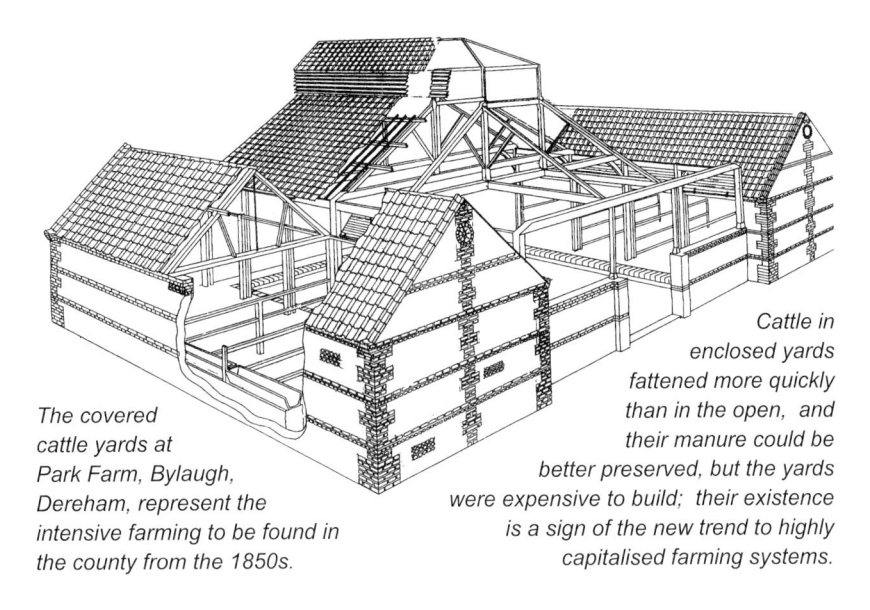

The covered cattle yards at Park Farm, Bylaugh, Dereham, represent the intensive farming to be found in the county from the 1850s.

Cattle in enclosed yards fattened more quickly than in the open, and their manure could be better preserved, but the yards were expensive to build; their existence is a sign of the new trend to highly capitalised farming systems.

The greatest developments in farm building were in the period traditionally associated with high farming. Increasingly intensive methods of livestock fattening influenced layouts as the principles of industrial design were taken up on farms. They throw light on the extent to which high farming ideas affected agricultural practice across the region and how much, and indeed how long, landowners were prepared to invest in such improvements after cereal prices began to fall in the 1870s.

The 1850s were the decade in which interest in farmstead design reached its height, first with the Royal Agricultural Society's farm building competition and then with a series of publications of farm building designs of which John Bailey Denton's *Farm Homesteads of England* is the most famous. The aim of all these plans was to create an efficient working unit where everything was arranged for the greatest convenience, manure could be conserved and cattle fattened quickly. This involved not only the housing of cattle in yards surrounded by shelter sheds, but also the building of ranges of loose boxes, often with a central feeding passage and turnip or cake houses at one end. Finally the open yard could be completely roofed over to provide protection for both animals and manure. Covered yards were regarded with some suspicion by Norfolk farmers because of the problems a lack of ventilation could cause for the well-being of the cattle, and few were built. However, one range survives south of Norwich at Swardestone and a further group at Winfarthing, with the finest example being those at Park Farm, Bylaugh, near Dereham, built for the extremely wealthy Evans Lombe family.

Estate records show expenditure on building increasing through the middle years of the century. At Holkham, building improvement costs rose to over £10,000 in many years in the middle of the century. This was over 25% of estate income from rents. There was hardly one of the 70 farms on the estate which did not have a set of new cattle sheds between 1840 and 1870. On the Walsingham estate, expenditure was high between 1859 and 1878, often over 10% of estate income, at between £500 and £1,000 a year. Between 1857 and 1869, Mr Preedy, the agent for the Le Strange estate at Hunstanton, presided over its improvement 'from one of the most dilapidated in the country to one that compared favourably with neighbouring properties'.

An area of major improvement where the landlord's leadership was particularly important was in the fens. The Hare estate covered about 10,500 acres near Downham Market, 2,000 of which were fen, regarded until the 1840s as very poor land which was let with the upland farms, 'incapable from its poverty of separate cultivation'. Between 1846 and 1849, the fens were drained and six new farms created. After 1860 many of the buildings of the upland farms were also rebuilt and incorporated features of industrial architecture such as sliding doors which could be left open without any danger of being blown off their hinges.

In contrast to the west, estates in the south of the county needed a great deal to be done to make them suitable to high farming systems. The buildings of the Duke of Norfolk's estates near Kenninghall were of wood or clay lump with thatched roofs and 'not adapted to the present mode of farming' as they lacked good cattle accommodation. The problems on the Earsham estate were similar. In 1864 they were described as 'totally dilapidated' and 'an entire reinstatement is absolutely necessary'. In both cases some

CHANGING AGRICULTURE

new brick and tiled buildings were built, often alongside older barns, whilst completely new well-laid-out farms were built at Earsham Hall and South Lopham. In sample areas across the county where an analysis of the layout of buildings on the tithe maps of the 1840s and the first edition 1:2500 maps of the 1880s has been made, it is clear that generally this was a period of rationalisation of buildings. In the 1840s half the farmsteads were of irregular layout, whilst by 1880 only 10% were irregular, more than half having an

PLAN OF NEW FARM-BUILDINGS ERECTED AT HILL FARM

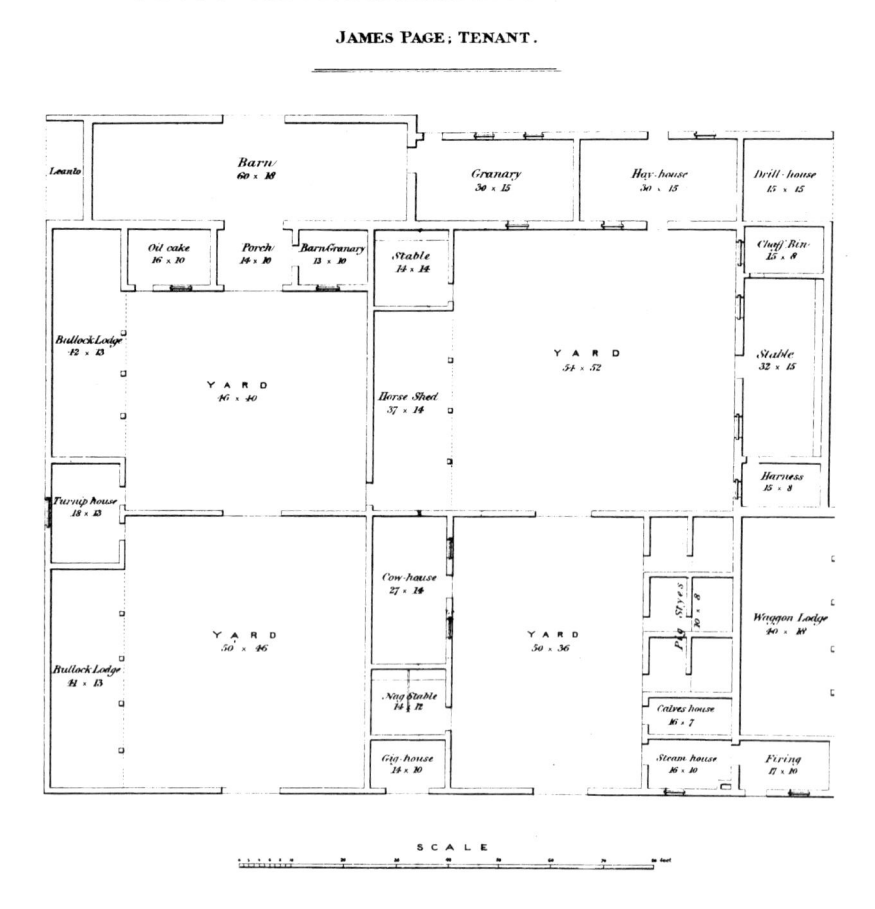

JAMES PAGE; TENANT.

The Hare estate at Stow Bardolph was building new farms as fenland was reclaimed. This plan shows a regular layout around yards, which was seen as the ideal by the 1850s.

HIGH FARMING

E or U plan with shelter sheds enclosing yards. When he wrote his report in 1858, Clare Sewell Read noted that there had been a great improvement in buildings. 'All over the county there are excellent new premises and generally the aspect of the old ones is decidedly improved.'

Underdraining

The importance of underdraining as part of improvement was stressed again and again by the commentators of the time. The agricultural writer and editor of the *Journal of the Royal Agricultural Society of England,* Philip Pusey, wrote that 'thorough draining is to the land as foundations are to a house' and between 1840 and 1855 over 10% of the articles in his *Journal* dealt with this topic. This interest was stimulated by the invention of a machine in the 1840s which could manufacture clay drain pipes. In 1846 drainage loan companies were established which could give long term loans for land improvement, but little of their money found its way into Norfolk. However, it is clear that

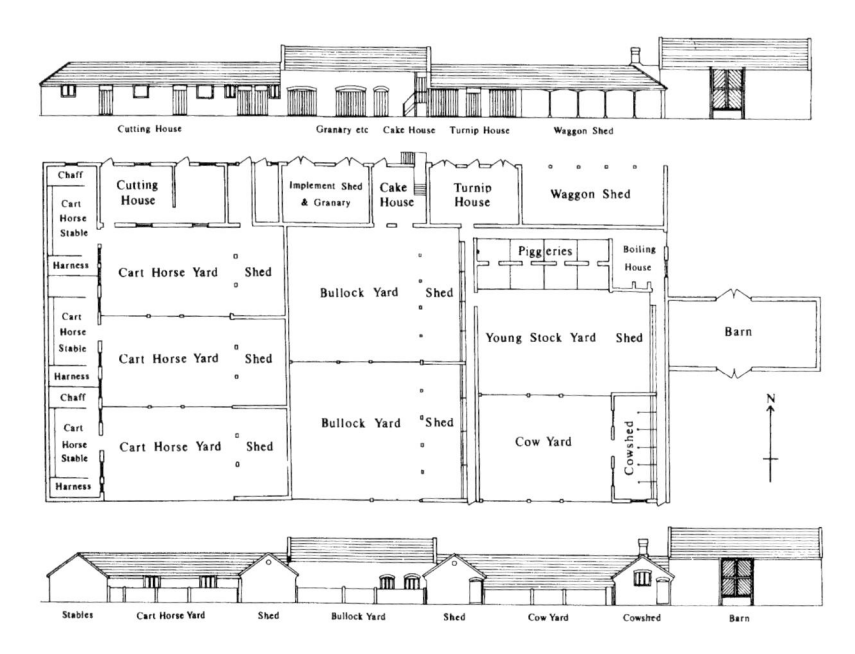

It was rare for farms to be completely rebuilt, but the buildings at Hall Farm, South Lopham, had become so derelict that they were replaced in 1860 by a carefully planned farmyard.

CHANGING AGRICULTURE

tile drainage was carried out on many estates without the help of loans. On the Holkham estate over £2,000 a year was spent on drainage throughout much of the 1850s. Drainage books survive for many of the heavy land farms on the estate showing the layout of the drains and listing new work and improvements to the systems. The Holkham agent, Mr Keary, recommended to the Duke of Norfolk that he should drain his south Norfolk farms on the same basis as at Holkham where the landlord was responsible for the work and then charged the tenants 5% *per annum* on the capital outlay.

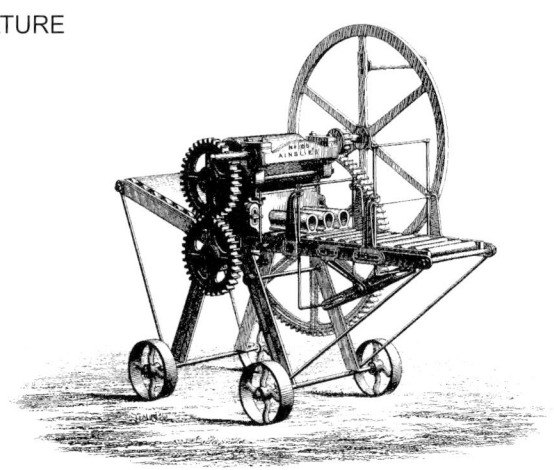

It was the invention of the drain-pipe making machine in the 1840s that enabled clay pipes to be mass-produced and much of the heavy land drained and thus more easily worked. These machines were illustrated in an agricultural encyclopaedia of 1855.

It is also true that, as we have seen, there had already been a considerable amount of drainage work in Norfolk using the old bush drainage method. This impermanent system could last up to 30 years in the heavier soils. A high proportion of the heavy Norfolk soils was owned by small estates or owner-occupiers who could not afford the expenditure involved in laying pipes and would have continued to dig bush drains. Other than underdraining and farm buildings, which were usually the responsibility of the landlord, the two main expenses for the farmer were in feed for his livestock and fertilisers for the land. We shall consider these two elements separately.

'Artificial' feeds and livestock production

Oil cake, a by-product of the rape and linseed oil extraction industry, was first introduced on British farms as a manure. Faden's map of 1797 shows an oil mill near Hillington, but more usually they were located in ports.

HIGH FARMING

There is mention of the purchase of 3,300 rape cakes in the accounts of the Home Farm at Holkham as early as 1732 and both William Marshall and Arthur Young reported several instances of its use. In 1771 Young reported that John Carr of Massingham was feeding oil cake to cattle. Its benefit was measured, not in its food value, but in the improved quality of the manure: 'the dung he raised from them was twice a beneficial as the cakes themselves spread upon the land'. The increase in its popularity is shown in its rising price from £3 to £6 a ton through the eighteenth century. The basic principles of the feeding systems which were to remain at the root of high farming were in place by 1800, but it was the years after 1820 that saw the most dramatic increase in the consumption of cake by both sheep and cattle. Bacon thought that the most important development of the late 1830s was the 'increased use of artificial feeds for fattening'. Like Young, he measured its value in the increased quality of the farmyard manure and the fact that it allowed more sheep to be grazed. The high farmer *par excellence* and a founder member of the Royal Agricultural Society was John Hudson of Castle Acre. With his father, he farmed 1,400 acres and in the 1830s he spent between £2,000 and £3,000 a year on oil cake and other artificial feeds. This allowed him to double the number of stock kept and so increase the output of the land by a third. There were of course the sceptics in the 1840s. Mr Overman of Weasenham used no oil cake, whilst Mr Burgess of Docking wrote, 'the expense is too great to pay a return on capital laid out; the only advantage is that the land is left in a high state of cultivation for the ensuing crops'.

The value of feeding cake was measured in terms of its enriching affect on manure. It was not until the arrival of the railways made transport to market so much easier that farmers were able to realise the full value of their stock. Instead of being subjected to a week-long walk to Smithfield on which cattle lost 28 pounds of weight, they could now reach London within the day with no loss of condition. Meat prices were also beginning to rise as urban living standards improved, and so farmers were gradually shifting towards a system in which cattle for market were playing a larger part. The number of cattle fattened in Norfolk rose substantially during the period and the number of cattle feed suppliers listed in Norfolk trade directories rose from 24 in 1858 to 54 by 1875.

As the number of cattle kept increased, a larger proportion were being bought in and home breeding declined. C. S. Read noted that few sheep and fewer cattle were bred in the region and it was rapidly becoming a fattening area. These animals needed to be yard-fed and new shelter

Farms could avoid the cost of covered yards for cattle by using loose boxes for fattening cattle. Arranged down each side of a central feeding passage and opening off the cattled yards, they first appeared on Norfolk farms about 1850; these are at Egmere Farm.

sheds and cattle yards were built. By the 1850s some cattle were being fattened up for market in loose boxes and these rows of boxes with a central feeding passage began to be built on the larger farms by the more progressive landowners such as Thomas William Coke on his new farmstead at Egmere Farm, near Walsingham.

The number of breeding sheep within the county also declined with the great ewe flocks confined to the light lands of Breckland which provided fast-maturing Norfolk-cross-Southdown lambs for the turnip fields elsewhere. Store lambs were valued for the manure they provided eating off the turnips before the barley crop. By the 1840s they too were being fed oil cake, usually between half a pound and a pound a day during the winter. This both improved the quality of the manure and allowed them to be folded more tightly so that more dung was produced.

Artificial manures

If the increased use of artificial feed intensified the livestock side of the farming enterprise, it was the use of artificial fertilisers which transformed arable husbandry. We have already seen that from the early eighteenth century oil cake was being spread as a fertiliser, and Arthur Young and his contemporaries listed a variety of organic waste products which were used experimentally in the fields. Bacon's report of 1844 contains details of various trials carried out by his correspondents using such products as nitrate of soda and potash, guano and saltpetre. However, it was farmyard manure, enriched by feeding oil cake, which was still the most favoured source of fertility.

By the 1830s the value of bone dust, particularly on the turnip crop, was becoming appreciated and imports trebled in the five years after 1832. There were bone mills at Kings Lynn and Yarmouth as well as on the river at Antingham, near North Walsham. By 1845 James Fison and Sons were bone and cake merchants in Thetford. The problem that soon became apparent was that bone dust was ineffective on alkaline soils. Only on acidic soils could the calcium phosphate be released. As a result of John Bennett Lawes' experiments between 1837 and 1842, it became clear that this reaction could be effected by treating the bone with sulphuric acid to produce superphosphates. At the same time it was shown that mineral phosphates rather than bones could be used and in 1842 Lawes was able to patent his method of producing superphosphates which created the fertiliser industry in Britain. The use of bonemeal declined with the introduction of superphosphates, but some farmers treated bones with acid themselves to create fertiliser. For instance in 1856 Mr Palmer of Eastmere Farm, Stanford, asked the Walsingham estate for a manure house in which he could dissolve bones. Manure manufacturers were established in the main ports and in 1847 Joseph Fison set up as a maltster and fertiliser manufacturer

It was the manufacture of fertilisers by firms such as Fisons at Thetford that allowed farmers to break away from the four-course rotation and produce heavier crops.

at Thetford, moving to a new vitriol and manure works just outside the town at Two Mile Bottom in 1853.

The impact of the new fertilisers within the region was clear by the time Clare Sewell Read wrote in 1858, when he noted that artificial manures had created the greatest revolution in farming over the previous 15 years, but he was sceptical about the value of many of the new products. In most cases farmers were out of their depth when it came to judging which was the best buy. The days when the greatest compliment that could be paid to a farmer was that he was a 'practical man' were past. Now it was scientific knowledge that was important. 'Our practical ignorance cannot be bliss unless it is pleasant to buy things at double their value and lose good crops into the bargain.'

As well as the chemical fertilisers there were imported natural products, such as guano from South America. This seagull manure was scraped off the coastal cliffs and sent to Britain in increasing quantities from the 1830s. By 1858 it was extensively used in west Norfolk for wheat, with a mixture of guano and superphosphate on barley. When the price of cereals fell, it was uneconomical to use and so imports dropped off rapidly after 1870.

In spite of the unreliability of many of the fertilisers on the market, C. S. Read credited them with being responsible for a spectacular increase in wheat of up to 25% in west Norfolk over 15 years. Formerly all the farmyard manure had been needed for the turnips, but now they were grown using artificials, leaving the manure for the wheat. The yield of barley had not increased since the 1840s, by which time as a result of increased stocking rates on the turnip fields, the barley crop, following on turnips in the rotation, had reached its optimum level. As well the large quantities of cattle feed purchased by John Hudson, he also spent between £800 and £1,000 a year on artificial manures such as superphosphates and guano for his Castle Acre farm. In 1851 the Holkham agent, Mr Keary was very impressed with what he saw there. 'That such lands produce this enormous bulk of roots and corn is truly astonishing and proves indisputably the high condition of the farm and the skill of the tenant.'

As a result of more and richer farmyard manure and the availability of fertilisers, it was possible to break away from the restrictions of the four-course rotation and grow more cereals. By 1858 the four-course was 'going out of favour in its native country'. Farmers were growing an extra crop of oats after wheat, and barley was sometimes sown into a wheat stubble. Two tenants of the Holkham estate, John Hudson and John Hastings, were asking for permission to depart from the terms of their leases and grow an extra grain crop in the 1860s, but by 1869 John Hudson had returned to the four-course, considering it 'better adapted to west Norfolk land'. Leases in general continued to stipulate a four-course rotation. As the prices for cattle were rising, landlords saw the value of high farming as allowing more cattle to be fattened for sale and more manure to be produced, which with the help of artificials would increase yields on existing acreages within the already established rotations, rather than growing larger acreages of cereals.

It is impossible to unravel the relative importance of the new feeds and fertilisers and land drainage to the jump in yields, particularly of wheat, that took place during the era of high farming, but the enormous variations in the quality of fertilisers might suggest that it was the increasing number of stock kept which was most important.

Mechanisation

A further much publicised element of high farming practice was mechanisation, but this, in contrast to other developments, made very slow progress in Norfolk. R. N. Bacon noted in his report of 1844 that 'the farmers of Norfolk have been guided more perhaps in the selection of their implements by the price than the excellence of their workmanship'. However, he did devote over 50 pages of his report to the subject of machinery. Again, C. S. Read, writing 15 years later, commented that although 'the progress in agricultural machinery had been very marked . . . Norfolk farmers appear rather slow in adopting new and improved implements'. This attitude, differing markedly from their enthusiasm for fertilisers, must reflect the fact that labour was still plentiful. The arrival of the railways in the 1840s reduced considerably the cost of coal and made steam power a more economic possibility; Burrells, who had been making seed drills and threshing machines from the 1800s, manufactured portable steam engines in Thetford from 1848. By 1858 there were said to be 'nearly as many engines as parishes' in Norfolk.

Traditional implements were also becoming more efficient so that by the 1840s even the heaviest land could be ploughed using two horses while one-horse ploughs were adequate on the lighter lands. By the 1840s levers had been added to scarifiers and other implements of tillage to regulate

CHANGING AGRICULTURE

The village sign at Corpusty is a gallows (or 'gallus') plough manufactured about 1864 by Ezra Cornish at the ironworks in the neighbouring village of Saxthorpe.

their depth in the soil. Machines that could be adapted to perform different functions at different seasons were also a valuable advance. Finally the development of portable threshing machines which could be hired from contractors was particularly important if smaller farmers were to benefit from mechanisation. Small farmers too could benefit from the more efficient ploughs and harrows which reduced the amount of horse power and therefore labour needed on the farm.

The major innovation of the 1850s was the reaping machine, welcomed by farmers as, unlike the thresher, it reduced the need for man-power in a season when it was in short supply. Two American machines, the MacCormick and the Hussey, were shown at the Great Exhibition in 1851 and were quickly taken up by British farmers. A few found their way into Norfolk in the 1850s. C. S. Read reported that they 'won golden opinions last harvest (1857) and the makers have in consequence received extensive orders from our county'.

Conclusion

Of the 6,500 farmers who filled in details of their farm size in the 1851 census only 63 farmed over 300 acres – the minimum size generally thought viable for high farming. A further 1,628 farmed between 100 and 300 acres and here too some elements of high farming may have been practical. By far the majority (4,239) farmed under 100 acres and their methods would hardly have been touched by the grand schemes publicised through the farming press. As C. S. Read wrote, 'Small farmers are not high farmers'. However, they would have benefited from the more efficient ploughs that were coming on the market and possibly hired a contractor with a threshing machine for a few days but they would not have bought much in the way of fertilisers and feeds. It is important to remember that for the majority of Norfolk's farmers little changed over the nineteenth century. They would have worked their land on a four-course rotation and relied on their own fodder to feed their stock and manure to keep their land in good heart. They would have replaced their bush drains every thirty years or so, but would not have been able to afford the new pipe drains. However, the spirit of confidence must have permeated through even to these smaller farmers who may well have tried out some cattle cake or 'artificials' from time to time, but who

Even after the introduction of the reaper binder, the crop had to be carted and stacked, and the crop that the binder had missed had to be raked up and bound into sheafs by hand. This group at Vale Farm, Stibbard, pictured about 1900, includes rakers and men with pitchforks to toss the sheafs onto the stack.

CHANGING AGRICULTURE

would no doubt have been very sceptical about the huge capital outlay of some of the most enthusiastic advocates of high farming.

Even for the large farmers, the profitability of high farming is difficult to assess. Some, such as John Hudson, made a fortune, but very few farmers were good book keepers and they rarely knew themselves what profits they were making. Figures for farms on the Breckland estates of Lord Walsingham between 1850 and 1869 suggest that outgoings of £1,000 a year on fertilisers were matched by the cost of labour and that little or no profit was made.

There can be no doubt that most mid-nineteenth century farmers believed that the key to increased productivity, and therefore prosperity, lay in science. Perhaps the optimistic philosophy of high farming is best summed up in the words of John Hastings of Longham, who when replying to R. N. Bacon's question, 'Has production reached its highest level?', replied 'This will develop in proportion as knowledge increases'. The high farming of mid-Victorian Norfolk was not just a question of fertilisers and cattle feeds but a state of mind exhibiting that confident faith in progress which was to be found throughout much of society at the time.

A sail cutter (reaping machine) of the kind introduced farms in the 1850s.

NORFOLK FARMING IN A CHANGING WORLD 1875–1914

Not all Norfolk farmers had the confidence of John Hastings. Clare Sewell Read farmed 800 acres at Honingham Thorpe, was an active member of the Royal Agricultural Society and a founder and first chairman of the Norfolk Chamber of Agriculture in 1865. He was author of reports on Oxfordshire and Buckinghamshire farming as well as those on Norfolk agriculture published in 1858 and 1881. In 1865 he was elected Liberal MP for Norfolk, thus becoming the first tenant farmer with a seat in the House of Commons, retaining it until 1880. Throughout the 1870s he was seen as the acknowledged political spokesman of the nation's farmers.

As early as 1858, when Read had reported that a quarter more wheat was grown in Norfolk than 15 years previously, he already feared that supplies of wheat from the New World were bound to increase and would undercut the price at which Norfolk's intensive farmers, with the high cost of inputs, could realistically afford to sell. By the late 1870s he thought that the limits of the gains possible through high farming had been reached. 'Beyond a certain point the increase in the crop is not proportionate to the increase of the manure supplied,' he said in Parliament in 1879.

Most farmers were taken by surprise when grain prices collapsed in the mid-1870s. In fact by 1870, the country already depended on imports for about a half of its wheat and 14% of its meat, but the fall in prices of British products did not really begin until the end of the American Civil War saw American and Canadian railways penetrating deep into the huge grain prairies to make export possible. As a result, between 1873 and 1896 prices dropped by an average of 40%, with the greatest falls in wheat. In January 1872 the price of wheat on Norwich market was 65s a quarter, but it was below 55s for the rest of the century, reaching its lowest at 25s in 1894.

This long-term decline was at first masked by a run of bad seasons from 1875, culminating in the wettest year on record in 1879 when the previous murmurings of the agricultural community swelled to a chorus of discontent as agricultural depression became a matter of national importance. Hardest hit were the grain farmers, especially those on marginal lands – both the Brecklands, which relied heavily on expensive inputs, and the heavy clays, which were difficult to work in the wet years. The price of good malting barley, for which Norfolk was famous, kept up better, with farmers selling to maltsters in London or the market towns. Farm malt kilns had been going out of business with the arrival of the railways which finally penetrated some of the more rural areas in the 1880s. A small farm maltings continued at Antingham Hall until the 1890s, but the agent to the Gunton estate advised against spending much on repairs. 'Would these malt houses be required by any other tenant? I should say no as country malt houses in the present day are of very little use or profit.' Poor quality barley, on the other hand, which often resulted from the difficulties of harvest in wet years, was unsaleable.

Barley has always been the most important cereal crop across Norfolk and the high price which good quality malting barley could command kept many farmers going through agricultural depression.

NORFOLK FARMING IN A CHANGING WORLD

Read's visit to America as part of the Richmond Commission on agriculture in 1879 convinced him that not only cereals but also livestock would suffer from competition. 'I really believe that grazing is the industry that will be ultimately injured most by American competition rather than grain,' he wrote in his report. He denounced science as a two-edged sword, was sceptical about chemical fertilisers which were easily washed away by rain and did not restore soil structure, and claimed that new machinery saved time and labour but not money. 'Practice may be slow to move out of her beaten track and science a little visionary, yet they exercise a healthy check on each other.' He therefore urged caution as far as 'high' farming was concerned and foresaw the need for the Norfolk farmer to diversify out of grain into, in the short term at least, livestock. On his own farm at Honingham he built a large covered yard in which to fatten his cattle more efficiently. However, even with these changes Read was pessimistic. As we have seen, he did not think livestock would be a long-term solution or that the change from arable to pasture which would accompany it was practical in Norfolk. Only by careful management and plenty of manure could successful grasslands be created. 'It seems a hopeless and almost impossible task to convert much of the arable land of East Anglia into permanent grass.' The soils were mostly too light and the climate too dry. In an outspoken criticism of the 'improvers' of the Napoleonic war years he wrote: 'It is a thousand pities that so much of our sheep walks, heaths and warrens were broken up years ago.' Where land was simply left to 'tumble down to grass', as had happened on a 'good deal of land' by 1888, foul grasses soon took hold. A final disadvantage of the change to grass was that less labour was needed and 'the population is almost banished'. But in spite of his pessimism, 70,000 more acres of permanent pasture had been created by 1880 than there had been thirty years before.

It is very difficult to know exactly how badly farmers were doing. Few kept detailed accounts and they were unlikely to publicise them anyway, except to emphasise their problems to a government inquiry. Read claimed to have made a loss on capital invested and found the only way to keep in business was to keep more animals and leave land in grass longer. He wrote in 1881, 'All is in favour of the man of capital who has plenty of money to stock his farm – but is terribly against the half-ruined tenant of less favoured land.' However, as he had predicted, livestock farmers had problems too; from 1864 the wool price was declining with competition from Australia and New Zealand. A decline in cereal prices meant that animal feed was cheaper, but the continued wet weather of 1880–81 resulted in foot rot in sheep and the loss of six million animals across England and Wales. This was fol-

CHANGING AGRICULTURE

lowed by outbreaks of cattle plague. Lord Lothian's agent at Blickling Hall felt it necessary to write to his employer in August 1885 to warn him of the problems being encountered by the tenants on his estates. Cattle plague was spreading across the county and a mutual assurance association for protection against loss had been set up in Norwich. The agent warned Lord Lothian to expect a diminished rental from his farms.

> The long-continued drought in the early part of the year was most injurious to every description of herbage and growing corn: of hay therefore there has been less generally than half a crop and of corn, both barley and wheat, there is a prospect of equal scarcity; added to which the rain has been so heavy and incessant as to put a stop to all harvest operations . . . the present state of things is most disastrous and the future most gloomy. Deficient crops, bad weather and the cattle disease all occurring together, will prove too much for many farmers.

There was little the Norfolk farmer could do to counteract the depression. The four-course system remained the basis of his system with some local variations. Some farmers on good soils were growing two grain crops in succession (generally barley after wheat), and elsewhere, particularly on light land, grass was left for more than one year and this became more general as grain prices dropped. Not surprisingly, in these circumstances, farms were difficult to let and many Breckland farms could not attract tenants and so were being managed by the landlord. In 1886 Lord Walsingham's agent wrote, 'There were times when I would have had scores of applications for such a farm . . . Few farmers have capital and those there are will not risk it at farming'. Farmers, he said, were living from hand to mouth. Because of this, in early September, bullocks were cheap to buy but 'they will be dearer after harvest when farmers thrash some corn and raise money with which to go to market'. Inevitably, rents went unpaid. 'The fact is farmers have little or no money. More than three quarters of the farmers in this and other counties are kept afloat by the bankers.'

As the situation became worse, Lord Walsingham was looking around for other solutions. He considered setting up a cooperative, letting the farm to a group of small tenants to farm, but his agent advised against it. 'The cooperative system would never answer . . . the best thing is to let farms, even at a very reduced rate rather than have them in hand. There are now more persons anxious to get out of farming than go in.' Even with rent reductions, Lord Walsingham was still left with several farms in hand

and was looking for various ways of diversifying, such as the growing of tobacco.

To reduce the amount spent on buying in feed, silage making was tried and the wet years of the early 1880s witnessed a short-lived enthusiasm for this experiment. The *Journal of the Royal Agricultural Society* reported on a 'silo and silage stack competition' in 1886 and in that year the silo of Mr Garrett Taylor of Whitlingham, near Norwich, was described and illustrated in the *Journal*. Lord Walsingham was a member of a government ensilage committee and a new silo was built at Merton in 1885 which would be 'one of the best if not *the* best in the kingdom'. In 1886 he wrote, 'we are finding the benefit of ensilage for the ewes and lambs now that the turnips are nearly done'. The Essex farmer and journalist P. Hunter Pringle took on Eastmere Farm, Stanford, in 1884 and asked Lord Walsingham if the money set aside to cover his cattle yards could be used for a silo instead. 'I intend as far as I can to be a successful farmer here,' he wrote, 'and I look to stock more than tillage to form the nucleus of my success.' The main problem was keeping the clamps air tight and for most farmers this was an expensive experiment that was soon abandoned, not to be tried again until the 1930s. By 1887 Hunter Pringle was less confident. He was taking in farm students but this was 'an uncertain side line' and he was looking for other employment, applying unsuccessfully for the post of secretary to the Royal Agricultural Society. He left Eastmere in 1889.

Agricultural depression brought with it the inevitable bankruptcies. In Breckland, Taylor of Tottington failed in 1888 amidst accusations to Lord Walsingham that he could not make a profit because his crops were eaten up by game, a claim that the agent described as a 'fearful untruth'. However, by the end of the century it was as sporting estates and plantations that Breckland was most valued. The costs of farming the lightest soils had become prohibitive and Read was proved correct in his views, first articulated in 1881, that 'it becomes every year more patent that a large area of sandy soil should never have been ploughed but have been allowed to remain in its original sheep walk or rabbit warren' (advice that the Holkham agent Ralph Cauldwell had given Lord Walsingham a century earlier). At the end of the century, Rider Haggard described the light lands around Swaffham which 'in the prosperous days commanded 7/6d an acre' as 'practically derelict'.

Further north, on the light chalk soils, the situation was slightly better. Lord Leicester was leaving land in grass for longer and keeping more sheep,

CHANGING AGRICULTURE

but he did not expect to be able to persuade many of his tenants to follow his example; most of them still believed that grain was the basis of farming prosperity. The well-to-do farmers of his Holkham estate were able to ride out the depression and pay their rents for a few years, relying on the income from their increasingly important livestock enterprises, but by the 1880s arrears were mounting and rents had to be reduced. In 1881 four farms had been vacated: 'one man has gone out because of declining health, one is dead, another thinks he had better get out before he has lost all he has got, and I'm afraid the fourth has lost all he had got.' Ten farms were vacated between 1880 and 1890 and a further 18 between 1890 and 1900. Many old-established names disappeared from the audit books. For the first time the estate had to advertise its farms. The Royal Commission on Agriculture reported: 'Perhaps Lord Leicester is an exceptional landlord, and no doubt it is a tradition on his estate that no good tenant is allowed to leave it, but even large reductions of between 42% and 56% have not always been able to keep tenants.' Farmers were not bringing up their sons to the land, 'as for £1,000 a man could be put into a profession whereas to farm he needed £3,000 which he stood a good chance of losing. Nearly all the people who hold small official positions in the district are farmers' sons or broken down farmers.'

The Hastings family had farmed on the Holkham estate at the outlying 580-acre farm of Longham Hall from at least 1757, and also owned land within the parish. The member of the family we know most about was John Sutton Hastings who held the farm from 1816, when he was 26 years old, until 1869. The beginning of his tenancy coincided with the reorganisation of the farm after an enclosure act of 1813. A new house was built and new buildings provided. To bring the farm into a high state of cultivation John Hastings had marled and drained the heavier soils. He described his system of farming for Richard Bacon's survey of 1844 and as we have already seen he was an optimistic man, interested in experimentation. 'I have grown experimentally almost every sort of wheat that I have seen.' He had changed to growing the more nutritious mangold rather than turnips, which tired of being grown frequently on the same ground. Here was a farm which, in spite of its poor soils, was being run on progressive lines by a farmer flexible enough to try new methods. During the 1850s, the number of livestock kept at Longham increased and new loose boxes were built for them. In the 1860s he gave up the four-course rotation for a few years, relying on artificials to provide the necessary fertility.

John Sutton Hastings died in 1869 at the age of 79 and his son, another

New cattle sheds were built in concrete at Godwick Farm in the 1880s, at a time when more cattle accommodation was provided by the Holkham estate to persuade tenants to stay. The experiment of building in concrete was short-lived, as brick and flint were both readily available.

John, took over. He continued to run the farm much as his father had, but relying more heavily on the livestock side of the enterprise to pay his rent. It was not until 1882 that he had problems paying. He was now ill, and his son, another John, gave notice that he intended to quit at Michaelmas and asked for a loan to keep him going until then. Tenants however were difficult to come by and he was persuaded to stay, allowing the rent to remain unpaid until the following February. In 1884 John Hastings senior died and his son was left with the difficult task of making the farm pay. The estate agreed to lend £500 for two years at 5% as well as providing a mortgage on some family property in neighbouring Gressenhall. In 1885 the rent at Longham was finally drastically reduced in line with that for many other farms on the estate, from £1,013 to £700. Even this could not be paid, however, and in 1887 he was £877 in arrears and again handed in his notice. The agent wrote to him, 'His lordship is extremely sorry to lose such an old and good tenant, but he feels it would be unfair to the rest of the tenantry to give further special remission in your case'. However, again he was persuaded to stay with a further rent reduction to £600. In March 1888, the agent was relieved to hear that 'after mature consideration you see your way to continue in the above occupation'. Further threats to quit were made

CHANGING AGRICULTURE

in 1891 and 1895 when the rent was further reduced to £477. John Hastings died in 1907 and the family finally left and so a 200-year association with the area was lost.

This story could be repeated on farms across the county. The belief that the depression was short-term meant that permanent reductions in rents were on the whole not made until the 1880s. Instead there were temporary remissions of up to 40%. Unfortunately for most tenants, the depression had lasted fifteen years before landowners agreed to renegotiate rents, often not before the end of long leases. Estates as far apart as Blickling in the north-east and Boyland Hall in the south were reducing rents by between a third and a half between 1880 and the end of the century. With these reductions, some of the burden of the problem was moved from the farmers to the landlords who were forced to retrench. The tenants themselves were in a much stronger position and were able to dictate their own terms to their landlords. However, as farms changed hands with increasing frequency old loyalties were severed and the relationship between landlord and tenant became a purely business one.

Although there were many changes in tenants in the 1880s and 90s there was no decrease in the number of farmers recorded in the censuses. By diversifying, farmers could make a living. At the time of the farm sale when the Hastings family left Longham in 1907, there was a breeding stable for London hackney horses on the farm. Fruit and vegetables, pigs and poultry, particularly for eggs, could all be produced profitably on small farms. The problem was that they were all products that could easily become overproduced with disastrous consequences for the farmer.

In 1886 the Royal Agricultural Society met in Norwich and two local farms entered the farm prize competition which was always part of the programme. The winner in the category of farms over 550 acres was Mr Garrett Taylor of Whitlingham Hall Farm who was experimenting with the use of town sewage as a manure. It was pumped to the highest point and then spread over 282 acres. No other fertiliser was needed for the corn and mangold crops and the judges for the competition reported that the cattle on the sewaged rye-grass pasture did better without cattle cake than those on the untreated pasture who were being fed cake as a supplement. Because the farm was on the outskirts of Norwich, Garrett Taylor was able to pursue one of the few branches of agriculture that remained profitable, that of dairying, and 100 red-polled dairy cows were kept. 'The milk is refrigerated, tinned and dispatched twice a day to Norwich by cart.' Twenty pigs consumed the waste from the dairy. After inspection of the books by the judges, the farm was said to be 'a financial success'.

The prize in the category 250 to 550 acres went to Mr Learner of Burgh-Next-Aylsham. It was his ability as a judge of good stock that enabled him to continue making a profit through the depression years. He bought about 150 young stock every autumn and fattened them on roots grown on the farm and on purchased cake. Mr Learner's 'judgment of the grazing qualities of his cattle, and his capability as a market-man of dealing with them' meant that there was 'not an inferior animal on the farm'.

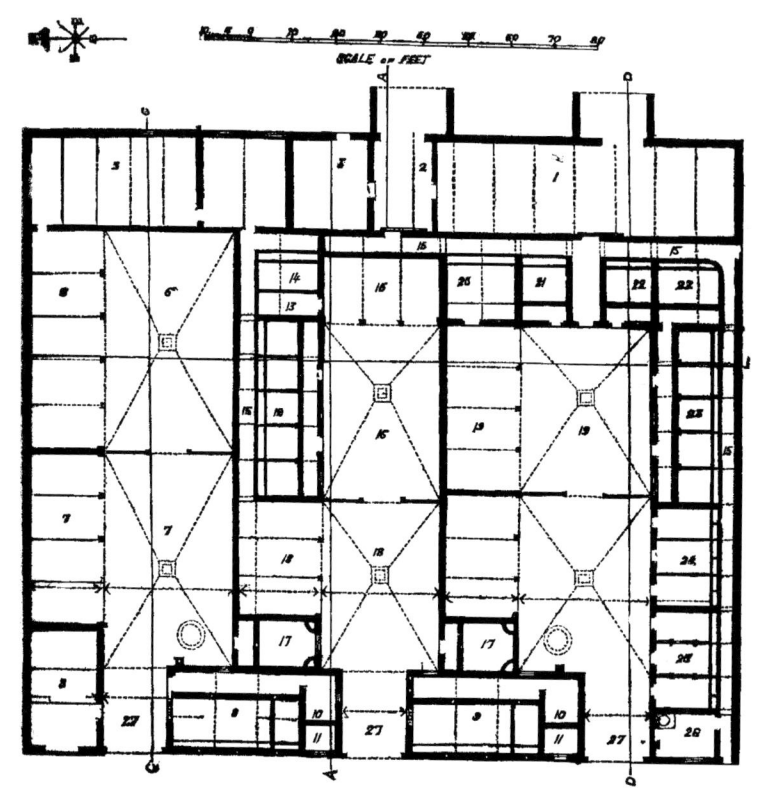

Bexwell Hall Farm was considered a truly innovative farmstead in 1886 when it won a prize in the Royal Agricultural Society farm prize competition. The buildings were arranged around yards with both shelter sheds and feeding stalls opening onto them. Passages behind all the stalls made it was easy to get food to the cattle.

97

CHANGING AGRICULTURE

We know little about the new tenants who were taking over. Some came from Scotland and brought with them Ayrshire cows with which they started up dairy herds and which gradually replaced the native red polls across the county. If their farms were near railway lines they were able to take advantage of the growing urban milk market. Mr Everington, who took over John Hudson's old farm in Castle Acre, was the son of a London merchant and a farm student of Clare Sewell Read. Mr Keith from Aberdeenshire took on Egmere in the 1880s and ran a highly mechanised farm where he was making a profit in the 1890s. 'Even the hedges were clipped by machine.'

It is clear that for farms to remain profitable rents had to be reduced considerably, often to the levels of the 1790s. Norfolk farmers had to be prepared to adapt, and here the ability and energy of the farmer was important. Good quality malting barley remained profitable, but costs had to be reduced, and this involved farming on a large scale and reducing labour. Livestock, particularly dairy cows, were likely to be the most remunerative, but it was expensive to buy the stock to establish a good herd. Sidelines such as the breeding of carriage horses could help, but again the foundation stock would be expensive. It was, as always, the large farmers with capital who could afford to make the necessary changes.

Changes to new crops could easily result in a glut. C. S. Read wrote in 1887, 'It is of little use telling farmers to alter their course of cropping, for no sooner do they substitute one produce for another than it becomes more depressed than the original crop.' The price of both barley and beef had dropped. Small farmers had taken to fruit growing, 'but the prospect last summer of acres of gooseberry and currant bushes and hundreds of plum trees all laden with fruit which would not pay for gathering was not encouraging.'

The real losers of the period of the Great Depression were the landowners, who saw their rentals drop by 50%. During the prosperous years many had invested much (up to 30%) of their increased rental in farm improvements such as buildings and underdraining, and now this investment was not showing the hoped-for return. Others, such as Lord Leicester, had been more canny, investing in the new industries through the expanding stock market. A list of investments in the archives at Holkham shows that up to the 1860s it was mostly local business, such as the Wells and Fakenham railway, that benefited, but from then on he was buying shares in railways world-wide so that by he end of the century his portfolio covered businesses in South and North America, India and Australia, and no doubt

NORFOLK FARMING IN A CHANGING WORLD

Bylaugh Hall was built between 1850 and 1852 to a design by Charles Barry the younger. However, agricultural fortunes declined and during World War 1 the Evans Lombe family decided to sell. The house fell into ruin.

these activities enabled the Holkham estate to weather the depression better than some. Others, such as the sixth Lord Walsingham, were not so lucky. Bad investments and an extravagant lifestyle which included entertaining the Prince of Wales, who had recently purchased Sandringham, led, by 1911, to insolvency.

Many landowners had taken out loans to finance house improvements which the dictates of fashion and the increasing complexities of country house life required. Some of these houses, such as Costessey Hall, have since been demolished and others, such as that at Bylaugh, are now in ruins, but the flamboyant house at Shadwell Park, transformed by the Buxtons in the 1840s, 50s and 60s following designs by the fashionable architects Edward Blore and S. S. Teulon, survives.

Debts which could no longer be serviced forced the smaller landowners to sell. Others, such as the Rolfes of Heacham, let the Hall and moved abroad, to Italy, where life was cheaper. Those with lands near Norwich, such as the Unthanks, or on the coast, such as Lord Suffield at Overstrand, could do well by selling for building, but others were forced to part with their land at prices that were as low as £20 an acre in the 1890s. The buyers

CHANGING AGRICULTURE

were often graziers and dairy farmers from the north of England or Scotland who had first arrived in the county as tenants. Because of the lack of capital, farm repairs were not carried out and any new building that was undertaken was in an effort to attract new tenants or to persuade existing ones to stay. Mr Rose reported on conditions on the Gunton estate in 1894. Not only were the buildings generally derelict but many of the farms were being let to non-resident tenants. Much of Bradfield, for instance, was let to Mr Ives, who neglected the buildings on outlying farms.

There was much discussion about the relative merits of large and small farms. Mr Rose thought the 'day of the large farm was over' and that a 180-acre holding in Roughton should be let to a resident tenant and not amalgamated with a neighbouring farm. As we have seen, Lord Walsingham was interested in the idea of the cooperative farm and the Liberal party championed the cause of small holdings. The cause was taken up by the Conservatives and the mostly ineffectual Small Holdings Act was passed in 1892. C. S. Read thought that good farming was synonymous with large holdings. He acknowledged that small farms were suited to the production of milk, pigs, poultry, fruit and vegetables and that these perishable goods were suffering less from steadily falling prices. There were still 20,000 farms of under 50 acres in Norfolk in 1880 and they certainly had some advantages. Less labour needed to be employed by the small family farm and it could survive on low returns. 'A small farmer, content with small profits, depending on the proceeds of garden and dairy produce, and commanding the labour of his family, may make both ends meet where a large

Mr John Senter of Vicarage Farm, North Elmham: a typical late-nineteenth-century Norfolk farmer.

NORFOLK FARMING IN A CHANGING WORLD

The Senter family outside Vicarage Farm, North Elmham, in the 1890s.

farmer becomes insolvent.' But the small farmer could well be worse off than a farm labourer and there was little chance to accumulate capital, and so in the long run he was always at a disadvantage. The large farmer on the other hand had access to capital, was likely to be better educated and could take advantage of economies of scale. However, it was the products of his farm, cereals, sheep and fat cattle, which were suffering most in the depression.

Prices reached their lowest in 1894, but thereafter there was a slow and steady rise. With rents adjusted down, agriculture began to become more prosperous. Mechanisation, particularly on the harvest fields, meant that labour costs had also gone down as a third of the labour force left the land between 1875 and 1900. A new generation of farmers who had survived the depression realised the need for agricultural education and experimentation. The Norfolk Chamber of Agriculture set up an Experiments Committee in 1885 and three of its members offered land for experimentation which it controlled until 1905. In that year its role was taken over by the Department of Agriculture in Cambridge and a permanent experimental station was set up at Jex Farm, Little Snoring, in 1908. The realisation that profits could only be made if standards were maintained is shown in the

CHANGING AGRICULTURE

interest shown in improved sheep and cattle breeding. The crossing of the old Norfolk Horn with the Southdown had produced a new breed of sheep to be known as the Suffolk and the Suffolk Sheep Society was founded in 1886. The red poll cow was a dual purpose animal originating in Suffolk and much favoured by the gentry. The Earl of Leicester kept a pedigree herd at Holkham and the Red Poll Cattle Society was founded in 1888. This activity is an indication of the more optimistic mood amongst livestock producers that was becoming apparent by the turn of the century.

The years up to 1914 saw a great shift in the balance of agricultural production. Good quality produce, particularly the malting barley for which Norfolk was famous, could still find a good market, but generally it was livestock which was most important. More land was put down to permanent grass, which increased from 214,479 acres in 1871 to 288,510 in 1891. Dairying was the major growth industry, but good quality fat stock sold well, especially now that there were railways to take animals to market. Fruit and vegetables, including potatoes, were becoming established in the fens.

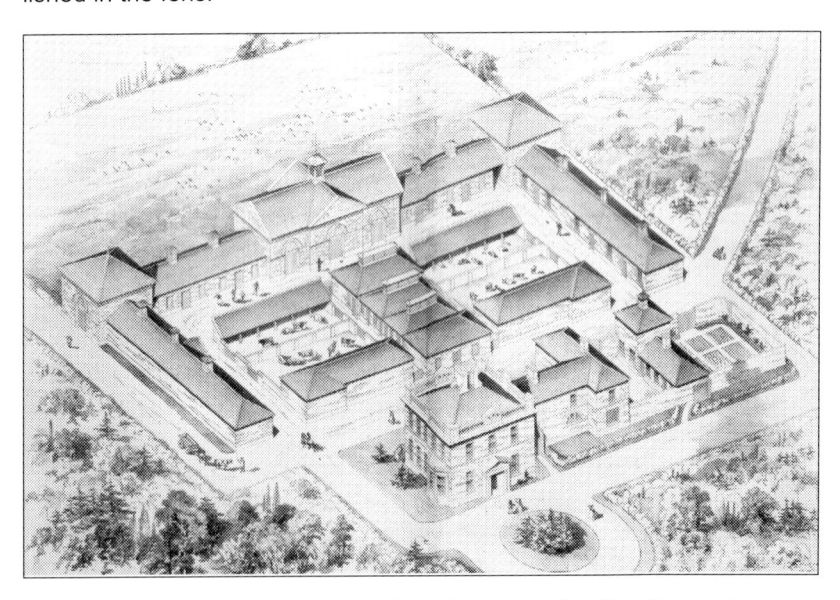

The Earl of Leicester was a prominent breeder of red poll cattle at his model farm in Holkham Park. It was a dual purpose breed used for both beef and milk production, but it was replaced by Ayrshires as Scottish farmers moved into the area at the end of the nineteenth century. (Plate 25, a model farm, Holkham drawing from A Series of Selected Designs for Country Residences . . . *by George Alfred Dean, 1867).*

NORFOLK FARMING IN A CHANGING WORLD

Perhaps the most significant change was the decline in the influence of the landowner who had lost most during the depression years. The landlord-tenant system, which C. S. Read had seen as central to British agriculture, could not survive the fall in rents and was finally broken by the great land sales after the end of the first world war. It was particularly the smaller owners who could not survive. Of the 32 estates of between 2,000 and 3,000 acres in 1880, only five retained a thousand acres in the hands of the original family a hundred years later, while of the 11 estates of over 10,000 acres in 1884, eight remained in the hands of the same family, albeit often with much reduced acreages. As estates were sold, their farms were mostly bought by sitting tenants and so the whole basis on which farming had developed over the previous 200 years changed. Nationally, the number of owner-occupier farms rose from 11% to 36% in the years immediately following the war. The old landlord–tenant relationship whereby the capital costs of farming had been divided between the landlord (who had provided the fixed investment) and the tenant (who provided the working capital) came to an end. Although this in theory should have given more freedom to the new owners on their own land, most were heavily indebted to the banks, and for much of the inter-war period farming suffered from a lack of capital.

It is only since the second world war, with the increasing availablity of grant-aid for improvement, that farming has returned to being the capital-intensive industry that it was during the prosperous century before the 1870s. The philosophies of the Victorian 'high' farmers have been embraced by a new generation and unimagined levels of production, mechanisation and economies of scale achieved. Now, however, we see a return of the notes of caution voiced by Clare Sewell Read in the 1860s, warning that chemical fertilisers could all too easily be washed out of the soil, that only farm yard manure could restore the soil structure, and that anyway the expense of 'high' farming was only worthwhile so long as there was a market for the product.

In the Middle Ages, Norfolk was the most prosperous and populous region of Britain, producing barley and wool for sale and supporting a thriving cloth industry. As industry moved away to the areas where water and steam power could replace manpower, its redundant work-force provided the labour for an intensive agriculture to feed the new towns, and so it became 'the cradle of the agricultural revolution'. So long as the home market for cereals remained secure, Norfolk agriculture was bound to prosper, with its large-scale, wealthy farmers continuing to adopt new techniques as they became available.

CHANGING AGRICULTURE

With increasing grain imports from the 1870s, agriculture was forced to adapt, and much of the earlier prosperity ebbed away. Two world wars forced the nation to recognise the value of self-sufficiency in food, and agriculture became increasingly tied to government and more recently European policy. This had led in recent years to the destruction of much of the evidence of that agricultural history. Hedges have been removed and redundant farm buildings demolished or converted beyond recognition. As farms get ever larger and smaller farmers are forced out of business and surpluses mount, we are being forced to re-assess the role of farming within the local and national economy. The countryside is still seen as a national asset and Norfolk still remains a very rural county with pockets of traditional landscape. In the past land was valued primarily for its potential to produce food. In the future, in some areas, this may have to take second place to recreation and wildlife. Norfolk was the first county to found a Naturalists Trust (in 1926) and also to implement an Environmentally Sensitive Area scheme (designed to protect the fragile environment of the Halvergate Marshes in the Broads) in 1987. It was here too that, in 1999, the most publicised anti-GM crop trial demonstrations took place. As Norfolk has led the way in the past, it is to be hoped that it will continue to respond to the changing national needs of the twenty-first century.

SOURCES

Books and articles

B. J. Armstrong, *A Norfolk Diary* (1949)
B. J. Armstrong, *Armstrong's Norfolk Diary: Further Passages* (1963)
R. N. Bacon, *The Report on the Agriculture of Norfolk* (1844)
P. Barnes, *Norfolk Landowners since 1880* (1993)
R. Brigden, *Victorian Farms* (1986)
D. Defoe, *A Tour thro' the Whole Island of Great Britain* vol 1 (1724)
J. R. Fisher, *Clare Sewell Read, 1826–1905* (1975)
H. M. Jenkins, 'Lodge Farm, Castle Acre, Norfolk' in *Journal of the Royal Agricultural Society of England* second series 5 (1869) pp. 460–74
N. Kent, *General View of the Agriculture of the County of Norfolk* (1796)
W. Marshall, *The Rural Economy of Norfolk* 2 vols (1797)
W. J. Moscrop, 'Report on the Farm Prize Competition in Norfolk and Suffolk' in *Journal of the Royal Agricultural Society of England* second series 22 (1886) pp. 566–664
M. Overton, 'The diffusion of agricultural innovation in early modern England: turnips and clover in Norfolk and Suffolk 1580–1740' in *Transactions of British Geographers,* new series 10 (1985) pp. 205–21
R. A. C. Parker, *Coke of Norfolk: a Financial and Agricultural Study* (1975)
A. J. Peacock, *Bread or Blood: a Study of the Agrarian Riots in East Anglia in 1816* (1965)
J. H. Plumb, 'Sir Robert Walpole and Norfolk husbandry' in *Economic History Review,* second series 5 (1952), pp. 86–9
C. S. Read, 'Recent improvements in Norfolk farming' in *Journal of the Royal Agricultural Society of England* 29 (1858) pp. 265–310

C. S. Read, 'Agriculture of Norfolk' in W. White, *History, Gazetteer and Directory of Norfolk* 4th ed. (1883)
N. Riches, *The Agricultural Revolution in Norfolk* (1937; Frank Cass reprint 1967)
H. Rider Haggard, *Rural England* vol. 2 (1902)
J. M. Rosenheim, *The Townshends of Raynham* (1989)
F. M. L. Thompson, 'The Second Agricultural Revolution' in *Economic History Review* 21 (1968) pp. 62–77
P. Wade-Martins (ed.), *An Historical Atlas of Norfolk* 2nd ed. (1994)
P. Wade-Martins, *Black Faces: a History of East Anglian Sheep Breeds* (1993)
S. Wade Martins, *A Great Estate at Work: Holkham and its Inhabitants in the Nineteenth Century* (1980)
S. Wade Martins, *Turnip Townshend, Statesman and Farmer* (1990)
S. Wade Martins & T. Williamson, *Roots of Change: Farming and the Landscape in East Anglia, c.1700–1870* British Agricultural History Society monograph (1999)
S. Wade Martins & T. Williamson (eds), *The Farming Journal of Randall Burroughes* Norfolk Record Society (1995)
A. Young, *The Farmer's Tour through the East of England* 4 vols (1771)
A. Young, 'The husbandry of Thomas William Coke' in *Annals of Agriculture* 2 (1784) p. 382
A. Young 'A week in Norfolk' in *Annals of Agriculture* 14 (1793) p. 456
A. Young, *General View of the Agriculture of the County of Norfolk* (1804)

Primary sources

Parliamentary Papers

1833 Select Committee on Agriculture, minutes of evidence
1834 Report into the administration and practical operation of the Poor Laws XXX Appendix (B1) Answers to Rural queries
1854 Agricultural statistics for Norfolk and Hampshire, collected by Sir John Walsham and Mr Hawley LXV
1881 Royal Commission on Agriculture XVII
1894 Royal Commission on Agriculture XVI
1896 Henry Rew's 'Report on Norfolk' XVII

SOURCES

Norfolk Record Office

Anmer estate records MC40
Bulwer-Long estate (Heydon) BUL
Earsham estate MEA
Hare estates (Stow Bardolph) HARE
LeStrange Estate (Hunstanton) Lestrange
Petre estate (Westwick and Buckenham Tofts) Petre boxes
Red Barn Farm, Snettisham, Kings Lynn Borough Archives af231
Walsingham estate (De Greys of Thompson and Merton) WLS

Private archive collections

Holkham Hall:
 Agricultural letter books
 Audit books
 F. Blaikie, 'Report on the states of T. W. Coke' (1816)
 William Keary's report on the estate (1851)
 General estate deeds
 Guardians' minutes
 Letter books

Raynham Hall:
 Townshend papers

Wolterton Hall:
 Archives of the Walpole estates of Wolterton and Mannington

INDEX

agricultural depression 61–2, 89–100
agricultural revolution 8, 10–11, 22, 62
Anmer 23
arable vs. pasture 91
Armstrong, Benjamin 46
Ashill 40
Aylsham 58
Bacon, Richard Noverre 64, 68, 69, 70, 75, 81, 82, 85, 88, 94
Balls, William 42
bankruptcy 61, 93, 99
banks 61, 92
barley 21, 26, 53, 84, 90, 98
barns 20, 21
Beevor, Thomas 49
Belhaven, Lord 31
Bexwell Hall Farm 97
Billing (tenant farmer at Weasenham) 49
Billingford 34–5, 46, 51, 69
Blaikie, Francis 37, 38, 45, 51–2
Blickling estates 20, 92, 96
Blofield 56
Blyth, Henry 64
Blythe, John 60
bone mills 83
books for farmers 11, 48–9
boulder clay 13
Bowells, John 21
Boyland Hall 96
Bradfield 100

Brampton Gurdon estate 37
Breckland 13, 25, 26; enclosure 40
Brettenham 25
brick buildings 21, 36, 55–6
broads 13–14
Brockdish 63, 64
Buckenham 40
Bulwer Long family (Heydon) 20
Burgess (farmer at Docking) 81
Burgh-next-Aylsham 97
Burlingham 63
Burnham 60
Burnham Deepdale 64
Burrells 85
Burroughes, Randall 46, 52, 55, 57, 63, 64, 65, 66
butchers 57
Buxton family 99
Bylaugh 76, 77, 99
cabbages 57
capital 51, 75–80, 94, 98, 101, 103
Carr, John (tenant farmer at Massingham) 33, 49, 57, 81
Castle Acre 61, 66–7, 68, 81, 84
cattle 21, 54–8, 59–60, 81–2; breeds 96–8, 102
cattle drovers 57
cattle fairs 55–7
cattle sheds and yards 36, 55–6, 76–7, 81–2, 95, 97
Cauldwell, Ralph 41, 93

INDEX

Cawston 17–19, 21, 39, 58
cereals 52–4, 84
chalk 11–13
Chambers (tenant farmer at Earsham) 72
chemical fertilisers 83–4
claylands 13, 14–16
clover 10, 16, 22, 26, 53
Coke, Edward 22
Coke, Thomas 33
Coke, Thomas William, 1st Earl of Leicester of Holkham 8, 32–9, 59, 82, 94; monument to 37, 38, 39, 54, 58–9
commons 15, 16, 22; enclosure 15, 33–5, 43
concrete building 95
Corpusty 39
Costessey 8
Costessey Hall 99
country houses 99
Coxford 24
Cranwich 26
Creake 24, 47
Cromer Ridge 13
crop rotation 10, 24, 37–8, 45, 53
Cullyer, John 49
dairy farming 15, 21, 48, 55, 96–8, 102
Davy, Humphry 49
Defoe, Daniel 8, 20–1
De Grey family (Merton) 40–1, 51, 61–2
De Grey, Thomas 31, 40–2
De Grey, William, 1st Lord Walsingham 40–1, 88, 92, 93
Dent, Peter 42
Denton, John Bailey 76
Dilham 48
diversification 93, 96
Docking 81
Docking (Cawston) 17, 39
Dorsey, Peter 48
Downham Market 64
drainage 10, 34, 75, 79–80; of marshland 13–14, 26–7, 77
Duffield, Mrs (Tottington small holder) 42
Dutch influence 28
Earle family (Heydon) 20
Earle, Erasmus (Heydon) 21

Earsham 15, 57, 62, 72, 77, 78
East Rudham 24, 66
Egmere 82, 98
Elveden 26
employment in agriculture 7, 62
enclosure 10, 14, 17–20, 24, 32–43; by act of parliament 40–3; of commons 15, 33–5, 43
examples mapped 17–19, 34–5
Evans Lombe family (Metton & Bylaugh) 77, 99
fairs 55–8
fallow year 10, 26, 53
farm buildings 36–7, 43, 47–8, 76–9; *see also* cattle sheds
farm houses 20, 21, 46–8, 74
farm size 43–4, 46, 86–7, 100, 103
farmers: education 46; mobility 46 social status 47–8, 74
Farrer (prospective tenant farmer) 51
Feltwell 27
fens 13–14, 26–7, 77, 102
fertilisers 75, 82–5, 91; *see also* manure
field size 52
Fincham 66
Fisons 82
flails *see* threshing
foldcourse system 22, 24, 40, 41, 58
food production 52
Freethorpe 56
Frost, Goddard 48
fruit farming 96, 98, 102
furniture 48
Fussell, G. E. 48
Gee (farmer near Norwich) 70
Godwick 95
grain prices 53, 60–1; rise in Napoleonic wars 24, 63; fall after wars end 35; collapse in 1870s 89–90
Grant, Robert 48
grass vs. arable 91
grazing 21, 24; marshland 27
guano 84
Gunton estate *see* Harbord family
Haggard, H. Rider 93
Halvergate Marshes 104
Harbord family (Gunton) 20, 99, 100

109

CHANGING AGRICULTURE

Hare estate 36, 77
Harleston 55
harvest 68, 69, 87
Hastings family)Longham) 94
Hastings, John 95
Hastings, John junior 95–6
Hastings, John Sutton 35, 84, 88, 94
Heacham 59, 99
heaths 17, 20, 22, 25, 34; reclamation 39
hedges 14–15, 24, 39, 59
Hempton Green 57
Hethel 49
Hethersett 63
Heydon 71
Heydon estate 20, 53, 55
'high' farming 54, 55, 74–88
Hill, Richard 21
Hillington 80
Hobart family (Blickling) 20
Holkham estate 22, 29, 32–9, 98–9; drainage 80; farm buildings 36–7, 55, 77; model farm 102; studied by historians 8; tenants 46, 49–51, 61, 84, 94; use of oil cake 81
Holkham leases 37–8
Holkham monument 37, 38, 39, 54, 58–9
Honingham Thorpe 89, 91
horse engines 71
Houghton estate 22, 29
Hoxne 57
Hudson, John 68, 81, 84, 87
Hunstanton estate 24, 77
Hunt, Rowland 61
improvement 9, 29–32, 40
industrialisation 7–8
insurance 92
Ives (tenant farmer at Bradfield) 100
Keary (agent at Holkham) 80, 84
Kenninghall 77
Kent, Nathaniel 37, 44, 48–9, 58, 63, 65, 67
Kerdiston 51
Kilverstone 25
Kipton Ash 58
labourers 62–73
landlord-tenant system 30–2, 45–6, 103
Langford 60

Langley estate 72
Learner (farmer at Burgh-next-Aylsham) 97
leases 30, 36, 37–8, 44–5; length 30, 36, 37–8, 44–5; stipulating crop rotation 24, 38, 53, 85; stipulating manual labour 72; stipulating marling 30, 36, 59
Lee, William 33
Leeds, John 46, 51, 54–5, 61–2, 69
Leicester, Lord see Coke
Le Strange family (Hunstanton) 22
Letton Hall estate 37
Levett, Thomas 42
Lincoln (tenant farmer) 59
literacy 28, 46, 48
Little Snoring 102
Longham 35, 88, 94–5
loose boxes 82
Lord, Robert 26
Lyall, John 48
machine breaking 72
malting 21, 90
mangold wurzels 57
manure 10; and cattle 55; guano 84; oil cake 82, 801; sewage 96; and sheep 22, 59
Margitson (agent at Earsham) 62
marl pits 24
marling 30, 33–4, 59
Marshall, William 21, 43–4, 49, 54–6, 57, 58, 63, 68, 70, 81
marshes 21
Marshland Smeeth 27
Massingham 33, 57, 81
Mattishall 58, 74
mechanisation 69–73, 85–6, 101
Mellet (tenant farmer at Dunton) 49
Merton 31, 93
Merton Hall estates 40–1
migration to urban areas 7
Moore, Tuttell 52
Neale, William 42
'Norfolk agriculture' 10–11, 30
Norfolk, Duke of 37, 77, 80
Norfolk Naturalists' Trust 104
North Elmham 100, 101
North Walsham 72
Norwich 57

INDEX

oats 84
oil cake 57, 75, 80–1, 82
open fields 10, 22, 25, 43; *see also* enclosure
Outwell 27
Overman (farmer at Weasenham) 81
Overton, Mark 21–2
Palmer (farmer at Stanford) 83
Parker, R. A. C. 33
parliamentary enclosure 40–3
pasture vs. arable 91
Peake (tenant at Tittleshall) 51
Petre, Lord 40
ploughing 67, 69
ploughs 67, 86
Plumstead 71
population growth 7, 64
Potter, Richard 24
Preedy (agent at Hunstanton) 77
Priestland, William (agent at Raynham) 24
Pringle, P. Hunter 93
probate inventories 21, 48
Prothero, R. E. 8
Purdy (tenant farmer at Castle Acre) 61
Pusey, Philip 75, 79
railways 81, 85, 102; investment in 98–9
raking 68
Raynbird, W. and H. 65
Raynham estate 22, 24, 29
Read, Clare Sewell 89, 91, 98; reports cited 75, 79, 81, 84, 85, 86, 87, 91, 93, 98, 100–1
reaping 69, 86, 88
Reepham 51
regions 11–27
rents 30–1, 60–2; during 1880s depression 92, 95–6, 98; on Hare estates 36; on Holkham estate 35
Riches, Naomi 8
Ringstead 65
Rolfe family (Heacham) 99
Roughton 100
round houses 71
Royal Agricultural College 75
Royal Agricultural Society: building competition 76; farm prize competition 96; motto 6, 9, 75

Rudham 24, 71
rye 26
St Faiths 55–7
Saxlingham 63
Saxthorpe 21
Scole 8
scythes 69
seasons and farm labour 67
seed drills 69–70
Senter family (North Elmham) 100–1
Senter, John 100
Sepping, Thomas 70
Sewell (tenant farmer at Thetford) 60
shackage 23–4
Shadwell Park 99
sheep 22–4, 26, 48, 57–9, 82, 91; breeds 58–9, 102
sheep shearings 38–9, 58
sheep walks 33, 59, 93
Shelfanger 16
Shotesham 63
silage 93
Sloley 74
small holdings 96, 98, 100
Smith, Adam 31
Smith, John 65
Snettisham 23, 49, 72
soil improvement *see* manure; marling
soil regions 11–27; map 12
sources for historiography 8
South Burlingham 20
South Creake 47, 71
South Lopham 78–9
sporting estates 93
Sprowston 63
stall feeding 21
Stanford 40, 41, 74, 83, 93
Starston 64
steam engines 71, 85
Stibbard 87
Stiffkey 24
Stow Bardolph 36, 77
strips 16, 34–5, 39
Sturston 40, 41, 75
Styleman, Nicholas 49
Suffield, Lord *see* Harbord family
superphosphates 83, 84
Swaffham 93
Swardestone 77

111

CHANGING AGRICULTURE

Swift, Jonathan 31
Swing riots 72
Syderstone 70
Taylor, Garrett 93, 96
tenant farmers 24, 28, 30–1, 46–52, 95–6
Terrington 27
Terrington St Clements 48
textile industry 8, 63
Thetford 40, 60
Thompson 40, 71
Thompson, F. M. L. 11
Thorpe Market 45
threshing 64, 70–3, 85–6
Tickler, Robert 24
tithes 8
Tittleshall 51, 52
Tivetshall 53
Tottington 40, 41–2, 59, 93
Townshend, Charles, 2nd Viscount ('Turnip Townshend') 22, 24, 26, 32
Tull, Jethro 69–70
Turner, John 49
turnips 10, 22, 24–5, 40; as break crop 53; as fodder 16, 21–2, 52, 55, 59, 60; labour required 69; and manure 84; stored in cattle sheds 55–6
unemployment 63, 66, 72–3
Unthank family 99
Upton-with-Fishley 56
Upwell 26

wages 63, 65–6, 68, 73
Walpole 27
Walpole family (Holkham) 20; estates 24
Walpole, Robert 22
Walsingham estate 77
Walsingham, Lord *see* De Grey
Ward (tenant at Warham) 51
Warham 49, 52
warrens 13, 25, 41, 93
Waterden 36
Weasenham 81
West Tofts 59
wheat 52, 54, 89; prices 60–1, 89–90
Whitlingham 93, 96
Wigful (patentee in Lynn) 71
Wiggins, John 36
Williamson, Tom 9
Wimbotsham 48
Winfarthing 77
Wolterton estate 20
woodland 15, 16
Woolpit 55
Worstead 8, 58, 63
Wretham 25
Wyatt, Samuel 47
Wymondham 46, 63
Young, Arthur 11, 30, 36, 48–51, 53, 63, 65, 66, 70, 71–2, 81, 82